AF754796

SAFETY IS **SEGURIDAD**

A Workshop Summary

Committee on Communicating Occupational Safety and Health Information to Spanish-speaking Workers

Committee on Earth Resources

Board on Earth Sciences and Resources

Division on Earth and Life Studies

NATIONAL RESEARCH COUNCIL
OF THE NATIONAL ACADEMIES

THE NATIONAL ACADEMIES PRESS
Washington, D.C.
www.nap.edu

THE NATIONAL ACADEMIES PRESS 500 Fifth Street, N.W. Washington, DC 20001

NOTICE: The project that is the subject of this workshop summary was approved by the Governing Board of the National Research Council, whose members are drawn from the councils of the National Academy of Sciences, the National Academy of Engineering, and the Institute of Medicine. The members of the committee responsible for the summary were chosen for their special competences and with regard for appropriate balance.

This study was supported by an agreement between the National Academy of Sciences and the U.S. Department of Health and Human Services, National Institute of Occupational Safety and Health. Any opinions, findings, conclusions, or recommendations expressed in this publication are those of the author(s) and do not necessarily reflect the views of the organizations or agencies that provided support for the project.

International Standard Book Number 0-309-08706-6

Additional copies of this summary are available from the National Academies Press, 500 Fifth Street, N.W., Lockbox 285, Washington, DC 20055; (800) 624-6242 or (202) 334-3313 (in the Washington metropolitan area); Internet, http://www.nap.edu

Copyright 2003 by the National Academy of Sciences. All rights reserved.

Printed in the United States of America

THE NATIONAL ACADEMIES

Advisers to the Nation on Science, Engineering, and Medicine

The **National Academy of Sciences** is a private, nonprofit, self-perpetuating society of distinguished scholars engaged in scientific and engineering research, dedicated to the furtherance of science and technology and to their use for the general welfare. Upon the authority of the charter granted to it by the Congress in 1863, the Academy has a mandate that requires it to advise the federal government on scientific and technical matters. Dr. Bruce M. Alberts is president of the National Academy of Sciences.

The **National Academy of Engineering** was established in 1964, under the charter of the National Academy of Sciences, as a parallel organization of outstanding engineers. It is autonomous in its administration and in the selection of its members, sharing with the National Academy of Sciences the responsibility for advising the federal government. The National Academy of Engineering also sponsors engineering programs aimed at meeting national needs, encourages education and research, and recognizes the superior achievements of engineers. Dr. Wm. A. Wulf is president of the National Academy of Engineering.

The **Institute of Medicine** was established in 1970 by the National Academy of Sciences to secure the services of eminent members of appropriate professions in the examination of policy matters pertaining to the health of the public. The Institute acts under the responsibility given to the National Academy of Sciences by its congressional charter to be an adviser to the federal government and, upon its own initiative, to identify issues of medical care, research, and education. Dr. Harvey V. Fineberg is president of the Institute of Medicine.

The **National Research Council** was organized by the National Academy of Sciences in 1916 to associate the broad community of science and technology with the Academy's purposes of furthering knowledge and advising the federal government. Functioning in accordance with general policies determined by the Academy, the Council has become the principal operating agency of both the National Academy of Sciences and the National Academy of Engineering in providing services to the government, the public, and the scientific and engineering communities. The Council is administered jointly by both Academies and the Institute of Medicine. Dr. Bruce M. Alberts and Dr. Wm. A. Wulf are chair and vice chair, respectively, of the National Research Council.

www.national-academies.org

COMMITTEE ON COMMUNICATING OCCUPATIONAL SAFETY AND HEALTH INFORMATION TO SPANISH-SPEAKING WORKERS

SUSAN C. SCRIMSHAW, *Chair*, University of Illinois at Chicago
ROBIN BAKER, University of California, Berkeley
JAMES PLATNER, Center to Protect Workers' Rights, Silver Spring, Maryland
LARRY SHAW-SALAZAR, IHI Environmental, Phoenix, Arizona

NRC Staff

TAMARA L. DICKINSON, Senior Program Officer, Board on Earth Sciences and Resources
TANJA PILZAK, Research Assistant, Board on Agriculture and Natural Resources
KAREN L. IMHOF, Senior Project Assistant, Board on Earth Sciences and Resources

COMMITTEE ON EARTH RESOURCES

SUSAN M. LANDON, *Chair*, Thomasson Partner Associates, Denver, Colorado
JAMES C. COBB, Kentucky Geological Survey, Lexington
VICKI COWART, Colorado Geological Survey, Denver
MURRAY W. HITZMAN, Colorado School of Mines, Golden
JAMES M. MCELFISH, JR., Environmental Law Institute, Washington, D.C.
JOHN MURPHY, University of Pittsburgh, Pennsylvania
DIANNE R. NIELSON, Utah Department of Environmental Quality, Salt Lake City
THOMAS J. O'NEIL, Cleveland-Cliffs, Inc., Cleveland, Ohio
DONALD PAUL, ChevronTexaco, San Francisco, California
RUSSELL STANDS-OVER-BULL, Arrow Creek Resources, Pryor, Montana
R. BRUCE TIPPIN, North Carolina State University, Asheville
MILTON H. WARD, Ward Resources, Inc., Tucson, Arizona
LAWRENCE P. WILDING, Texas A&M University, College Station
PHILLIP MICHAEL WRIGHT, Idaho National Engineering and Environmental Laboratory, Idaho Falls

NRC Staff

TAMARA L. DICKINSON, Senior Program Officer, Board on Earth Sciences and Resources
KERI H. MOORE, Program Officer, Board on Earth Sciences and Resources
KAREN IMHOF, Senior Project Assistant, Board on Earth Sciences and Resources

BOARD ON EARTH SCIENCES AND RESOURCES

RAYMOND JEANLOZ, *Chair,* University of California, Berkeley
JILL BANFIELD, University of California, Berkeley
STEVEN R. BOHLEN, Joint Oceanographic Institutions, Washington, D.C.
VICKI COWART, Colorado Geological Survey, Denver
DAVID L. DILCHER, University of Florida, Gainesville
ADAM M. DZIEWONSKI, Harvard University, Cambridge, Massachusetts
WILLIAM L. GRAF, Arizona State University, Tempe
RHEA GRAHAM, New Mexico Interstate Stream Commission, Albuquerque
GEORGE M. HORNBERGER, University of Virginia, Charlottesville
DIANNE R. NIELSON, Utah Department of Environmental Quality, Salt Lake City
MARK SCHAEFER, NatureServe, Arlington, Virginia
BILLIE L. TURNER II, Clark University, Worcester, Massachusetts
THOMAS J. WILBANKS, Oak Ridge National Laboratory, Tennessee

NRC Staff

ANTHONY R. DE SOUZA, Director
TAMARA L. DICKINSON, Senior Program Officer
DAVID A. FEARY, Senior Program Officer
ANNE M. LINN, Senior Program Officer
PAUL M. CUTLER, Program Officer
KRISTEN L. KRAPF, Program Officer
KERI H. MOORE, Program Officer
LISA M. VANDEMARK, Program Officer
YVONNE P. FORSBERGH, Research Assistant
MONICA R. LIPSCOMB, Research Assistant
EILEEN M. McTAGUE, Research Assistant
VERNA J. BOWEN, Administrative Associate
JENNIFER T. ESTEP, Administrative Associate
RADHIKA S. CHARI, Senior Project Assistant
KAREN L. IMHOF, Senior Project Assistant
SHANNON L. RUDDY, Senior Project Assistant
TERESIA K. WILMORE, Project Assistant
WINFIELD SWANSON, Editorial Consultant

Acknowledgments

This workshop summary has been reviewed in draft form by individuals chosen for their diverse perspectives and technical expertise, in accordance with procedures approved by the National Research Council's Report Review Committee. The purpose of this independent review is to provide candid and critical comments that will assist the institution in making its published summary as sound as possible and to ensure that the summary meets institutional standards for objectivity, evidence, and responsiveness to the study charge. The review comments and draft manuscript remain confidential to protect the integrity of the deliberative process. We wish to thank the following individuals for their review of this summary:

Edward Kissam, Aquirre International
Raja Ramani, Pennsylvania State University
Jackie Nowell, United Food and Commercial Workers International Union

Although the reviewers listed above have provided many constructive comments and suggestions, they did not see the final draft of the summary before its release. The review of this summary was overseen by Dr. Eula Bingham, University of Cincinnati, Ohio. Appointed by the National Research Council, she was responsible for making certain that an independent examination of this summary was carried out in accordance with institutional procedures and that all review comments were carefully considered. Responsibility for the final content of this summary rests entirely with the authoring committee and the institution.

Contents

Executive Summary

The growing presence of Spanish-speaking workers and employers in the United States and the unprecedented 12-percent increase in the overall rate of workplace fatalities among Hispanic workers in 2000 highlights the need to better communicate occupational safety and health information in Spanish to both employees and employers. To address this need, the National Institute for Occupational Safety and Health (NIOSH) is preparing a strategy for developing and disseminating Spanish-language occupational safety and health educational and technical material. To gather information necessary to create this strategic plan, the National Research Council (NRC) was asked to host a workshop to

(1) identify the most pressing occupational safety and health issues faced by Spanish-speaking workers and employers in the United States;
(2) examine how NIOSH can best meet the informational needs of the occupational safety and health community to address effectively the safety and health issues faced by Spanish-speaking workers and employers in the United States; and
(3) identify potential partnerships for NIOSH in reaching Spanish-speaking workers, their employers, and others.

To address the charge, the NRC established a committee and commissioned five white papers to set the stage for a workshop that was held on May 29-30, 2002, in San Diego, California. This workshop summary is a synopsis of the presentations and discussions at the workshop and is intended as input to the NIOSH strategic planning in this area. It does not contain any conclusions and recommendations. The conclusions and recommendations in the white papers represent the views of the authors and not necessarily those of the committee or the NRC.

Workshop participants discussed several priorities for NIOSH to consider in targeting health education campaigns, public health interventions, and occupational safety and health information to best address the elevated rate occupational injuries and fatalities among Hispanics. These priorities include:

(1) workers and employers who speak and write little or no English;
(2) recent immigrants rather than established populations;
(3) workers with low literacy levels in both English and Spanish; and
(4) workers with high-risk occupations and industry sectors.

Based on fatality rates, these high-risk industry sectors include agriculture, construction, food processing, and health care.

As described in the NIOSH National Occupational Research Agenda (NIOSH, 1996) priorities related to organization of work, understanding the barriers and context within which public health interventions can be effective should be a priority in targeting Spanish-language materials, as it is in targeting English-language public health interventions. In addition to language differences that can be addressed by translation, recognition of cultural differences may modify the nature of an effective public health intervention, even when addressing the same occupational health hazard in the same industry sector as an English-language counterpart. The National Occupational Research Agenda noted that effective interventions should extend beyond simple translation to address diversity within the Hispanic workforce and differences from the non-Hispanic workforce. Existing data to target, prioritize, and provide metrics for evaluation of the effectiveness of interventions is not currently limiting initial public health interventions, but the data have considerable weaknesses. Workshop participants thought that NIOSH, the Bureau of Labor Statistics, and the Occupational Safety and Health Administration (OSHA) should develop major long-term initiatives to improve public datasets, but that several high-priority Hispanic target populations can be distinguished with relative confidence using current data.

While it is clear from workshop discussions that occupational safety and health resources for Spanish-speaking workers are needed, it is less whether that there are adequate materials that fit these needs. Many different domestic and international sources of Spanish-language materials were identified. Workshop participants identified a variety of problems regarding existing materials. There was general agreement that there is a need to collect, evaluate, and disseminate Spanish-language resources on an ongoing basis. Workshop participants agreed that the quality of existing material is varied and some of it is often poor. To address the issue of quality of material, workshop participants urged that evaluation standards for materials be implemented, including an evaluation of existing materials. They suggested that a national clearinghouse (ideally, web based) of materials judged to be of good quality be established with some information on best situations for use practices.

The participants did not think that it was adequate to simply collect, evaluate, and disseminate existing materials. Development of new materials to fill current gaps would also be of great value. The language must be appropriate to the educational level of the target audience. The material should pinpoint behaviors that are key to safety, and focus on these. It is essential to develop both content and delivery modes for messages with consumer input.

With respect to delivery mode and approaches, the workshop participants agreed that multiple modes of delivery should be considered, not just print media. Workshop participants agreed that communication materials and strategies must be in Spanish, but that language alone is not enough for adequate communication. It is essential that messages be delivered in culturally appropriate ways both in terms of content and approaches. Discussion during the workshop focused on providing resources for those who are best able to access high-risk Spanish-speaking workers and to support their efforts.

There was also considerable discussion and varied opinions about the best approach to translating material into Spanish. There was agreement that direct translation of materials, warning labels, signs, etc. usually does not convey the correct meaning, and that adaptation is essential at the very least.

The workshop participants discussed how to prioritize which information gaps to fill and believed that the priority should not be on translation of technical health and safety documents. Rather, emphasis would be better placed on developing materials as part of a strategic initiative to reach Spanish-speaking workers, their employers, and their communities with practical information that can assist in preventing workplace injury and illness. Well-developed educational materials by themselves do not always assure that the intended audience will be reached. It is important that materials be used in the context of a well-planned educational intervention. Workshop participants thought that it would be important to have a solid evaluation process.

The workshop participants appreciated the opportunity provided by NIOSH for them to come together to discuss these issues. The diversity of the participants was noted. Participants included representatives from government agencies, community organizations, academic research centers, employers, outreach workers, and union members. Discussions were open, honest, and productive. There was agreement on the importance of the problem, particularly in relation to the numbers of workers affected, the risks inherent both in common occupations for Latinos, in workers with little or no command of English, and in the lack of power to affect change or ask for their rights (e.g., workers' compensation, OSHA and immigration rights). The power differential between workers and employers is particularly strong because of the lack of legal status of many workers, even when employers recruit from across the border. The need to protect all workers, as reflected in the initiative of NIOSH in convening this conference, was acknowledged by the participants.

1

Introduction

Approximately 32.8 million persons of Hispanic descent live in the United States, half of whom were born outside the United States (Therrien and Ramirez, 2000). By the year 2050, it is expected that Hispanics will constitute more than 25 percent of the total U.S. population and approximately 15 percent of the U.S. labor force. These estimates and the fact that 90 percent of Hispanic American men and 60 percent of Hispanic American women participate in the U.S. workforce strongly suggest a need for occupational safety and health information in Spanish.

According to the National Immigration Forum, in 1990 more than 40 percent of new immigrants in the United States reported that they did not speak English "well." This statistic dropped below 25 percent for the immigrant populations who have lived in the United States for 10 years (Rodriguez, 1999). The 2000 Census reported that 27.3 percent of Hispanics, 25 years and older, had less than a 9th grade education (U.S. Census Bureau, 2001a). Educational attainment in 1994 of Hispanics of Mexican descent was significantly less than non-Hispanic whites. For a population of 25 years and older, 33 percent of Mexican born Hispanics in the United States had less than a 9th grade education, compared to 4.5 percent of non-Hispanic whites (U.S. Bureau of Census, 2001b). According to the Bureau of Labor Statistics, in 2000, work-related fatalities of U.S. Hispanic workers increased by 12 percent overall; in construction the increase was more than 24 percent. Since 1992, when the Bureau of Labor Statistics began accumulating statistics on Hispanic workers, work-related fatalities in this group has increased more than 50 percent.

The growing presence of Spanish-speaking workers and employers in the United States and the unprecedented 12-percent increase in the overall rate of workplace fatalities among Hispanic workers in 2000 highlights the need to better communicate occupational safety and health information in Spanish to both employees and employers. To address this need the National Institute for Occupational Safety and Health (NIOSH) is preparing a strategy for developing and disseminating Spanish-language occupational safety and health educational and technical material. To gather information necessary to create this strategic plan the National Research Council (NRC) was asked to host a workshop to:

(1) identify the most pressing occupational safety and health issues faced by Spanish-speaking workers and employers in the United States;
(2) examine how NIOSH can best meet the informational needs of the occupational safety and health community to address effectively the safety and health issues faced by Spanish-speaking workers and employers in the United States; and
(3) identify potential partnerships for NIOSH in reaching Spanish-speaking workers, their employers, and others.

To address the charge, the NRC established the Committee on Communicating Occupational Safety and Health Information to Spanish-Speaking Workers and Employers in the United States to undertake this study. The committee consists of four experts from academia, industry, and labor with expertise in construction and building, industrial safety and health, occupational health education, mine safety and agriculture. Brief biographies of the committee members appear in Appendix A. The committee commissioned five white papers (see Appendices D-H) and organized a workshop on May 29-30, in San Diego, California. The white papers were intended to set the stage for the workshop, and do not necessarily represent the views of the committee or the NRC. The workshop included participants from industry, academia, community organizations, and government with expertise in health and safety aspects of agriculture, mining, manufacturing, and service organizations (see Appendix B). The white paper authors made presentations and these were followed by discussion among the participants (see Appendix C).

This workshop summary is a synopsis of the presentations and discussions at the workshop. It does not contain any conclusions and recommendations. The conclusions and recommendations in the white papers represent the views of the authors and not necessarily those of the committee or the NRC. It is intended as input to the NIOSH strategic planning in this area. Chapter 2 discusses the available information and identifies information gaps regarding risks and adverse events for Latino workers. Chapter 3 examines the available health and safety training resource materials for Latino workers, especially for those with little or no English capabilities; in particular, it discusses issues of the linguistic and cultural appropriateness of materials. Chapter 4 considers issues surrounding the assessment of existing materials and the development of new materials. Chapter 5 discusses the various means of conveying information to Spanish-speaking workers, again focusing on cultural appropriateness and ways of maximizing understanding. Chapter 6 summarizes the discussion in the prior chapters and presents some overarching issues raised by the workshop attendees.

Workshop participants agreed that the highest priority for efforts to improve safety and health for this population should be on communication with Spanish workers who speak, read, and understand little or no English. Within this category there are differences in vocabulary by country of origin and differences in dialect within the country of origin. Some of these differences may extend to terms important for worker safety, such as names of equipment or tools. Some of the efforts to develop language-appropriate safety communications will need to take this into account. As will be discussed later, cultural variations in perceptions of risk and the need for protection are also important, as is the use of culturally appropriate ways to communicate safety information.

2

Hispanic/Latino Occupational Health and Safety: Available Information and Information Gaps

National census and household surveys such as the Current Population Survey, which over-sample[1] for minorities including Hispanics, provide critical demographic data for improving our understanding of shifts in the Hispanic population. In 2000, there were 35.3 million Hispanics in the United States, a 58-percent increase over the 22.4 million recorded in the 1990 census. This compares with a population increase of less than 9 percent among non-Hispanics during the same time period (U.S. Census Bureau, 2001). By 2050 it is projected that one out of every four persons in the United States will characterize themselves as Hispanic, compared with about one in eight today. While extremely valuable for tracking changes in our population, these datasets face a variety of challenges in their interpretation and in considering priorities for future efforts. Workshop discussion related to the white paper prepared by Scott Richardson et al. (see Appendix D) included several concerns about interpretation of Census data in driving development of information and educational outreach. They noted that the census methodology results in differential undercounts of Hispanics and the correlates of undercount are those most closely linked to risk for occupational injury. Therefore, the census population profile underestimates the size of the at-risk population. However, these data clearly document the rapid growth of the Hispanic worker and employer populations in the United States and demonstrate the need to consider health education and communication strategies that can more effectively reach these communities. Such demographic characteristics as location, education, country of origin, and language spoken at home are valuable tools in developing materials and public health interventions to serve at-risk populations more effectively. The adequacy of existing datasets and their potential use in assessing the impacts of future interventions on Hispanic worker occupational injury and illness were discussed. Key aspects of that discussion follow.

CENSUS PARAMETERS

Census methodologies do not always capture the detail needed for the types of analyses which are useful in assessing a specific population, as for this report. The section below discusses some of the data and data constraints in the most recent census.

Population Classifications

Recent reports from the Institute of Medicine (NRC, 2000, 2002) and the Centers for Disease Control and Prevention (CDC, 1993; Hahn, 1992, 1999; Hahn and Stroup, 1994) have

[1] Sampling is based on calculations of the number of randomly selected people needed to represent a population for the data analyses planned. At times, it is necessary to include more members of a specific group in the sample in order to have enough members of that group to be able to draw valid conclusions about that group from the data.

criticized conventional categories used to classify populations in the United States. Indeed, even the Office of Management and Budget has noted that

> these classifications should not be interpreted as being scientific or anthropological in nature, nor should they be viewed as determinants of eligibility for participation in any Federal program. They have been developed in response to needs expressed by both the executive branch and the Congress to provide for the collection and use of compatible, non-duplicated, exchangeable racial and ethnic data by Federal agencies.[2]

The workshop attendees agreed that there are multiple subcategories of Hispanics/Latinos. These include countries of origin from Spain to Mexico to Argentina, and length of residency from recent arrivals to multiple generations in the United States These groups and subgroups do not constitute a "racial" or even an "ethnic" group. As described in the sources mentioned above, there are variations in language, behavior, and other cultural aspects within and among these groups, as the groups and the individuals in them have different cultural experiences or cultural processes that shape and influence them (NRC, 2002). The workshop participants noted the change that occurred between the 1990 and 2000 censuses to allow designation of multiple racial groups may affect comparisons with non-Hispanic black and non-Hispanic white populations but should not affect the fraction of Hispanics of any race compared with non-Hispanics of any race.

The workshop participants agreed that a sensible public health agenda would give priority to addressing high-risk groups of workers. While absolute numbers of workers in various occupations or industry sectors should be given due consideration, groups with high fatality rates were deemed the highest-priority target populations. Defining such priorities using existing datasets is challenging. Using the Census of Fatal Occupational Injuries combined with denominator employment information from the Current Population Survey, it is possible to calculate fatality rates of Hispanics and non-Hispanics in various occupations and industry sectors. Non-fatal injury rates among Hispanics by occupation or industry can also be calculated. This provides valuable targeting information for defining high-priority occupations and industries with elevated Hispanic fatality or injury rates. During the workshop several cautions when using such data to target health education initiatives were discussed. This included concerns about the statistical validity of fatality rates in very small populations. If there are only two Hispanic elephant trainers and one dies in a given year, the calculated rate may be very high but may not reflect real relative risk of the occupation or be statistically meaningful. Similarly, if there are no fatalities in a given year, this may not accurately reflect risk. Therefore, strictly targeting the highest-risk or highest-relative-risk occupations or industry sectors could lead to misallocation of effort. The workshop participants agreed that it would be more appropriate to focus priorities on larger high-risk occupations or industry sectors where sample sizes are adequate to provide reliable results.

Self-Employed

The self-employed are exempt from most occupational safety and health regulations, including fatality, injury, and illness data collection and reporting requirements. As a result the Occupational Safety and Health Administration (OSHA) reports far fewer occupational fatalities than the Bureau of Labor Statistics' Census of Fatal Occupational Injuries, since the Census of

[2] See <http://www.nih.gov /od/ocpl/resources/ombclearance/directive15.pdf>. Date accessed November 5, 2002.

Fatal Occupational Injuries includes the self-employed and public sector workers. Small establishments (under 10 employees) have been excluded from reporting injuries. This may skew Hispanic injury and fatality data somewhat, although exactly how is unclear. Because barriers to entry into the labor market are likely to be lower with small establishments than large ones, this may disproportionately affect Hispanic employers and employees. Reportedly, individual workers are increasingly hired as self-employed subcontractors rather than as employees to avoid withholding of payroll taxes. This excludes them from protections such as workers' compensation insurance as well as occupational safety and health regulations. This is likely to be most common in workforces with limited economic opportunities or limited knowledge of U.S. and state legal protections. Although true self-employment can provide a path to economic opportunity, contract labor may lack such opportunities and yet be classified as self-employed in these datasets. There is no current research to verify that Hispanics are at increased risk of these sorts of employment practices. Such misclassifications of employment would skew reported injury and employment rates to the extent that such practices exist. Workshop participants agreed that further research to better characterize such organization of work factors and their impact on injury and illness reporting as well as actual injury rates would be valuable.

Personal services such as domestic help, home health care, and similar occupations present unique outreach challenges, since there are no concentrations of similarly at-risk employees. Self-employment and at-home businesses are a significant fraction of total employment that presents health education challenges and are likely to be undercounted in existing national datasets.

Industrial and Occupational Classifications

It should be noted that while the mining sector as a whole is identified as a high-risk sector for Hispanic fatalities, the majority of these Hispanic fatalities are in fact in the subcategory of petroleum exploration and production. Petroleum and gas exploration and production are covered under the Occupational Safety and Health Administration's general industry regulations (or similar state plans), rather than under safety and health regulations administered by the Mine Safety and Health Administration. Neither the U.S. Standard Industrial Classification system nor the North American Industrial Classification system nor the Standard Occupation Classification system (see Sidebar 1), which code occupations, use definitions that align with legislative and regulatory boundaries. The Mine Safety and Health Administration regulations may apply to a worker during the execution of some tasks, while OSHA regulations may apply while performing other tasks. Similarly there are different regulatory boundaries that affect occupational safety and health under the Railway Labor Act, Longshore and Harborworkers Act, Jones Act (maritime), and others. Employers in the same U.S. Standard Industrial Classification system or North American Industrial Classification system may have employees who work under a variety of occupational safety and health regulatory schemes. For example, an electrician may work under OSHA construction regulations, Mine Safety and Health Administration regulations, OSHA general industry regulations, Harborworkers Act, various state OSHA plans, or have no regulatory protection at all if the electrician is self-employed or works in the public sector in a state without an OSHA-approved state plan. The industry and occupation classification systems do not consider these legislative or regulatory boundaries, so to the extent that health education interventions address workers and employers in a given sector it may be important to communicate more than one regulatory scheme. This is also an important consideration in policy analysis research considering possible impacts on public health of legislative or regulatory actions or employer organizational changes. This presents challenges in defining relevant comparisons with national data.

SIDEBAR 1
Industry and Occupation Classification Systems

The Standard Occupational Classification system is the U.S. federal government's standard occupational classification system. An occupation is the kind of work a person does to earn a living. The Standard Occupational Classification system classifies workers according to the work performed into categories for the purpose of collecting and analyzing data. Each broad occupation category includes detailed occupation(s) requiring similar job duties, skills, education, and experience.

The Standard Industrial Classification codes were developed to facilitate the collection and analysis of census data. An industry is the type of activity at a person's place of work. The Standard Industrial Classification codes classify business establishments by the types of products or services they make available. Establishments engaged in the same activity, whatever their size or type of ownership, are assigned the same code.

In 1997 the North American Industry Classification system replaced the Standard Industrial Classification. This revision not only provides for newer industries but also reorganizes the categories according to production or process. The North American Industry Classification system is used for statistical reporting of all economic activities of the United States, Canada, and Mexico.

SOURCE: U.S. Census Bureau at <http://www.census.gov/hhes/www/ioindex/faqs.html#Q2>, date accessed: November 5, 2002.

The coding systems for both industry type and occupation are in the process of changing in order to achieve comparable classification systems across Mexico, the United States, and Canada and to improve comparability with the International Standard Industrial Classification system. The North American Industrial Classification system will replace the U.S. Standard Industrial Classification system and the Standard Occupational system will be adopted in 2003 for the Current Population Survey. Early versions of these systems were used in the 1997 Economic Census. These coding changes will result in shifts in populations in some categories. Categories with the same name under the U.S. Standard Industrial Classification system and North American Industrial Classification system may in fact have different contents under the new system due to significant differences in how establishments are classified (Walker and Murphy, 2001). For example, the construction sector will include approximately 3 percent more employees under the North American Industrial Classification system (Hiles, 2001) by incorporating construction laborers and helpers, landscaping, rental of construction equipment, and management occupations such as land subdivision, which were not included in construction under the U.S. Standard Industrial Classification. Residential remodelers is also a new listing in the 2002 version of the North American Industrial Classification system that is likely to have significant Hispanic employment but not have a comparable category in historical data. Exactly how this change in the coding systems will affect the reporting of Hispanic fatality rates by industry and occupation is unclear. Interpretation of trends and changes from historical rates may be difficult to interpret based on these national data sources during the transition period. Whether observed fatality or injury rate changes are the result of coding changes or actual changes in Hispanic fatality rates in targeted occupations or industries will be difficult to assess using national datasets. Workshop participants thought this should be considered in developing evaluation mechanisms for assessing public health interventions among Hispanic employers and employees.

Most industrial classification and occupation codes are by necessity broad enough to include diverse tasks, diverse levels of risk, and diverse work processes. In some cases these classifications may have inadequate detail to target particular high-risk tasks or processes.

Identification of a high-risk task, high-risk exposure to a hazardous material, or high-risk process cannot necessarily be used to define an exposed or at-risk population of businesses or employees using existing datasets. For example, laborers and helpers and non-construction may include a wide range of tasks and practices that present diverse risks of occupational injury or fatality. Public health interventions targeting specific tasks may have to incorporate evaluation mechanisms independent of national or state datasets in order to assess impacts. In another example, the work of cleaning a poultry or meat slaughterhouse falls under Building Cleaning and Maintenance Services, which includes office building cleaning services and janitorial services. The rate of injury for this industrial classification code is low because of the overwhelming number of cleaners in relatively safe, non-hazardous jobs compared to the slaughterhouses, which have very different exposures and risks. The cleaning of slaughterhouses is very dangerous and primarily employs Hispanic, non-English-speaking, new immigrants.

Temporary Employment Agencies

New entrants into the U.S. labor market are likely to be observed in employment sectors and settings with low barriers to entry. These include sectors with direct foreign recruitment, temporary employment agencies, casual or informal sector labor, day labor, and contract labor. While precarious employment often provides low barriers to new labor market entrants, it also makes demographic data collection, communication of health and safety risks, and dissemination of public health information particularly challenging. Similarly such industry structures are likely to result in increased part-time work, seasonal work, frequent new employers, and multiple simultaneous employers. Among small employers and the self-employed these factors may result in short-lived businesses, multiple sources of income (including self-employment), and intermittent contracts or business income. This sector of the Hispanic population offers unique public health intervention and health communication opportunities.

Temporary agencies and personnel services are a rapidly growing industry sector While employees of such services perform diverse tasks in occupations from construction and manufacturing to office clerical and janitorial services, these are all reported in a single industry category based on the principal activity of the establishment. It is, therefore, not clear which occupations are represented within this industry classification. The workshop participants were concerned that workers are reported in this single industrial classification and yet may perform work in any industry sector from custodial services and clerical to construction trades. There is further concern that Hispanics may disproportionately participate in the informal cash economy, which may not be accurately reflected in national statistics.

UNDERREPORTING

These datasets face serious challenges of underreporting among Hispanics in both the census (McKay, 1992) and in the Current Population Survey, particularly with mobile or migrant populations, populations without telephones (Current Population Survey is a telephone survey) or stable residences, and illegal immigrants who may not respond to surveys for fear of being identified. Illegal and low-visibility housing units, where the recently arrived, poorest, and limited-English Latino immigrants live, are disproportionately missed during mail surveys as well. There is a strong indication of underreporting of injuries among Hispanics reflected in the disparity between increasing fatality rates and higher than average occupational fatality rates combined with lower than average reported lost work time injury rates (see Appendix D; Center to Protect Workers' Rights, 2002).

The workshop participants discussed the issues of underreporting of injuries and undersampling and agreed that despite the remarkable changes documented in the formal statistics, there is a significant group of Hispanics that are not adequately sampled. However, because of the multiple data sources used in the Census of Fatal Occupational Injuries, the expectation is that recorded Hispanic occupational fatalities are relatively complete and are more accurately characterized than basic demographic shifts, occupational injuries, or occupational illnesses.

Workshop participants generally agreed that economic, cultural, and personal pressures were commonly exerted on employees to limit reporting of workplace injuries. Low-income Hispanics in precarious employment settings may be particularly susceptible to such pressures when they face economic uncertainty. Reporting of injuries results in direct costs to the employer in terms of Workers' Compensation costs and insurance experience modification ratings that affect future insurance premiums. Research to quantify the extent of injury underreporting and to define the impacts of economic and cultural incentives and disincentives would allow more effective dissemination of Occupational Safety and Health Administration information using injury data. Given reporting concerns for injuries and illness, current targeting of high-risk occupations and industries can best be achieved using occupational fatality data.

Workshop participants believed that immigration status is likely to have a significant impact on injury reporting behavior. Illegal workers who are recent immigrants may be unfamiliar with injury reporting practices and state Workers' Compensation systems, or may not want to risk dismissal as a result of any behavior that may not be viewed positively by the employer or supervisor. Illegal immigrants are likely to be in the most economically precarious situations making them responsive to any reporting disincentives. In addition, illegal immigrants may be unaware of the OSHA provisions and procedures for reporting unsafe conditions.

Workers in low-skill jobs where there is a ready supply of replacement workers may be particularly responsive to pressures to underreport injuries and illness. This may also influence the workers' willingness to perform tasks, question high-risk practices, and report violations.

Workshop participants thought that employer attitudes and the impact of direct discrimination should be the subject of further research. It may be that Hispanic worker injuries are less likely to be reported based on a variety of factors that may impact employer and supervisory attitudes on various worksites. To what extent can higher relative risk among Hispanics be explained by differences in occupation, age, education, and other demographic characteristics?

For low-income workers in precarious employment there is significant economic pressure to work even when injured or ill. Paid sick leave is unlikely to exist in such settings, and workers may be unaware of Workers' Compensation systems. Even if they are aware of Workers' Compensation systems, the systems may require multiple days out of work before any compensation is provided. Although failure to report injuries may result in long-term economic damage due to later medical complications or the result of repeated cumulative injuries or exposures, workers in economically fragile positions may focus largely on immediate income to meet immediate life needs. Research on such low-income workers is limited and presents researchers with serious challenges compared to research in stable employment settings.

Although workers involved in the informal setting (e.g., handywork, gardening, construction, construction cleanup, auto repair, gleaning fields) could conceivably be included in household surveys such as the Current Population Survey, they may be less likely to report such employment. Employers who answer this survey would be very unlikely to be included in the Economic census or other employer survey. Workers who are paid cash without withholding are not eligible for unemployment insurance payments, Workers' Compensation payments, health insurance payments of medical expenditures, or social security disability payments. These and similar economic disincentives are likely to affect reporting of occupational injuries. A quantitative assessment of the U.S. informal economy (i.e., economic transactions that avoid existing tax structures) would likely show that recent and illegal immigrants of any ethnicity are disproportionately involved in this cash economy.

Although the Current Population Survey allows the comparison of union and non-union workers, it is likely that economic incentives to report injuries and illness and undersampling in household surveys are quite different between these groups. Unions may offer valuable public health outreach capabilities for Hispanic workers in some sectors. Union status and other organization of work factors are important research variables in evaluating the effectiveness of public health interventions to prevent occupational injury and illness.

Undersampling is likely to be more severe in a mobile workforce. Typically the Current Population Survey requires a defined address (not a shared hotel room) to administer the survey. As mobility increases, including regular migration as in agricultural workers, it would be expected that undersampling would increase in any structured survey. This is a major concern in construction and agriculture, which are industry sectors with the highest fraction of Hispanic employment.

The Current Population Survey is a telephone survey that depends on having not only an address but also an active residential telephone (not a pay phone or hotel phone). This is likely to lead to undersampling the lowest income and most mobile sectors of the workforce. The Current Population Survey is administered in Spanish as well as English.

ENFORCEMENT DATA

OSHA datasets such as the integrated management information system citation data, occupational fatality investigation data, and injury-reporting surveys provide potentially valuable additional datasets, although these datasets do not exist solely for research purposes. Injury-reporting surveys involve submission of OSHA injury logs from a sample of approximately 50,000 establishments (excluding construction, government, and the self-employed). With the exception of recent changes in the OSHA 170 fatality investigation report form, these datasets do not typically include ethnicity, race, or language information. Analysis is difficult when this information is included as part of some narrative components of these datasets.

In terms of regulatory enforcement as an intervention to prevent injury and disease, workshop participants thought that further research is needed to assess the awareness of Hispanic employers and employees regarding OSHA requirements and penalties, and whether they constitute a significant incentive for improved practices. While OSHA provides many materials in Spanish and has a hiring priority to expand the number of Spanish-speaking compliance officers, it is unclear how effective these initiatives are at informing the Hispanic employer and employee communities. More effective coordination and sharing of Spanish-language documents across OSHA and among various state plans would be valuable, as would research evaluating the effectiveness of these materials at reaching various segments of the Hispanic community. Awareness of injury record keeping requirements may also affect the quality of data collected from Hispanic employers as a group. It was also noted at the workshop that OSHA spends most of its resources on large industries in heavily unionized workplaces. However, the greatest risk is with small businesses in non-unionized environments. OSHA interventions could be better targeted by focusing on the high-risk sectors.

SURVEILLANCE DATA

Serious challenges are faced by surveillance efforts, such as the NIOSH-supported Adult Blood Lead Survey and other state disease registries, state Workers' Compensation research, and injury and fatality data. There are several initiatives that could make it illegal to collect race and ethnicity data, and this may make continuity of data for future comparisons difficult. NIOSH-supported efforts such as the Adult Blood Lead Survey for adult blood lead concentrations, OSHA's new initiatives to add ethnicity and primary language questions to the OSHA Form 170 for fatality investigations, and similar long-term surveillance of injury and disease, which are critical for establishing baselines and tracking the impact of future public health interventions. Current trends within NIOSH and Center for Disease Control and Prevention (CDC) to convert cooperative agreements for state surveillance programs into contracts may create obstacles to recruiting additional state participants in existing surveillance programs or initiating new programs. Harmonization of core data across state boundaries is critical in characterizing illness or injury rates among ethnic or racial minorities where the small number of cases in some states may not allow reliable analysis and may raise confidentiality and privacy concerns.

Note that for the purposes of the Current Population Survey, both ethnicity (Hispanic, non-Hispanic) and race (e.g., black, white, Asian, Native American) are independently categorized. This requires caution when comparing Hispanic rates with black or white since these categories overlap. Comparisons with non-Hispanic whites or non-Hispanic blacks or with population averages are required for clarity. Further, a response of Hispanic, Chicano, Mexican American, or Latino does not necessarily equate with Spanish as a primary language or the language spoken in the home; rather, some fraction of self-designated Hispanics speak Spanish as the primary language in their homes. Another important issue is that there are households where the primary language spoken at home is an indigenous one and Spanish is the second language. Workers in these households may have limited ability to understand even orally presented occupational health and safety information in Spanish. This is an important issue in regions where there are higher levels of immigration of indigenous origin workers (e.g., California, Oregon, Florida) but not so important in other areas.

DATA PRIORITIES

The general view of the workshop participants was that for most high-risk activities, industry sectors and occupations the demographic and fatality data is adequate, and public health professionals know enough about the problems (e.g., causes of injury) and effective controls to provide the basis for developing interventions. Data on occupational illnesses remains inadequate not only for Hispanic workers but for all workers. Data on non-fatal injuries, although they are likely to be impacted by a variety of factors as outlined above, may provide the basis for useful comparisons in narrowly defined research initiatives and for prioritizing the most serious injuries where underreporting is expected to be less severe than for minor injuries. Use of these data is constrained by the fact that injuries are also likely to be underreported. The influence of social, economic, and cultural factors on reporting behaviors is critical before existing injury data can be effectively interpreted.

In the opinion of workshop participants, research that further quantifies and characterizes specific risks is not the principal barrier to preventing occupational injury and disease among Hispanic workers. Rather, research should focus on the effectiveness of specific field interventions with the goal of national or targeted dissemination of those interventions that are found to be efficacious in pilot projects, combined with research to evaluate the effectiveness on the health and safety of targeted populations. How can we reduce occupational injury and illness among Hispanics, starting with known hazards, among high-risk groups of employers and employees? Improved communication and improved knowledge of the organization of work, economics of occupational injury, and work culture is clearly an aspect of such a research and public health agenda.

There was concern at the workshop that research on safety culture, organization and sociology of work, and the economics of injury may be perceived as blaming Hispanic victims. Despite the potential difficulties, the workshop participants encouraged NIOSH to expand research on the organization of work and industry structures that are faced by Hispanic workers and employers. Participants also encouraged the development of experimental or quasi-experimental approaches in order to define and evaluate effective interventions to improve awareness, change organizational incentives and disincentives, and change high-risk behaviors.

A major concern of the workshop participants was the lack of appropriate long-term funding mechanisms for such interventions. Current NIOSH research funding, both internally and extramurally, focus on one- to three-year awards. The exceptions are a few five-year cooperative agreements. The current trend within NIOSH and the Centers for Disease Control and Prevention is reportedly to eliminate cooperative agreements in favor of shorter-term grants and contracts. This may not provide realistic funding mechanisms for developing, piloting, national or regional dissemination, and evaluation of public health interventions to prevent fatalities, injuries, and illness among Hispanic workers. Hispanic employers may be hesitant to allocate their limited resources to implement injury and illness prevention programs that have not been demonstrated to be effective in such field research. Mechanisms for planning, managing, and funding for applied intervention research on a 5- to 10-year time frame is critical. Public health interventions focusing on changing individual behaviors such as tobacco smoking and seatbelt use have required long-term perspectives, and there is every indication that organizations may be even slower to change. It is essential that those individuals be influenced and implement organizational change. Workshop participants agreed that such long-term funding mechanisms as cooperative research agreements are crucial as an administrative priority for more effectively conducting field research on evaluating interventions to prevent injuries and fatalities.

FUTURE DATA NEEDS

A recent addition of questions on principal language to the OSHA 170 fatality investigation form should provide some additional information for evaluating the extent to which communication problems may contribute to fatal injuries and development of injury prevention interventions.

Workshop participants agreed that improved occupational illness data is critical. This might be developed through various pilot initiatives targeting states, geographic areas, or high-risk groups. Priority could be given to at least one pilot program targeting a population that includes a significant Hispanic workforce. Improved injury data and research to further define individual and employer incentives and disincentives to report injuries are critical to understanding apparent disparities between Hispanic fatality and injury rates.

Marketing campaigns for disseminating public health information and interventions have been developed for such topics as prevention and cessation of tobacco smoking and seatbelt use. The occupational health community has limited knowledge of these interventions and how they might be adapted to fatality or injury prevention interventions targeting Hispanic employers or employees. During workshop discussions participants noted that multi-disciplinary research initiatives related to social marketing or models of effective intervention dissemination mechanisms are a critical area for future research. Data structure design and determination of reasonable metrics for defining changes from baseline present major challenges and may drive changes in the current datasets.

Workshop participants were not aware of any exposure databases that distinguish Hispanic workers. It is possible that existing exposure databases do not include ethnicity. Worksite data can also be problematic, since it is often poorly defined when dealing with some of the multi-employer worksites, multi-site employers, and transient contract employers.

Similar problems exist with task data. Workshop participants could not identify task inventories to compare Hispanic and non-Hispanic workers. For example, it would be of interest to know whether actual work practices are different among Mexican-trained workers who are recent immigrants. Particularly with very-low-wage workers, task inventories may be different with a higher fraction of manual and hand tool work, and fewer capital-intensive processes.

A significant concern raised by the workshop participants was youth employment (under 18 years old) when it places young workers at unacceptable levels of risk and limits their opportunities for education and future employment. Recent NIOSH recommendations on Department of Labor Hazardous Orders define allowable employment for youthful workers in the United States. These may be of particular value to Hispanic youth who may face high-risk work processes with reduced benefit from training and supervision that is often provided only in English.

Other areas where participants identified data gaps included longitudinal data on new immigrants, information on the informal economy, and workers in the cash economy.

SUMMARY OF INTERVENTION PRIORITIES

Participants agreed on several priorities for NIOSH to consider in targeting health education campaigns, public health interventions, and occupational safety and health information to best address the elevated rate of Hispanic occupational fatalities. These priorities include:

(1) workers and employers who speak and write little or no English;
(2) recent immigrants as opposed to established populations;
(3) workers with low literacy levels in both English and Spanish; and
(4) workers with high-risk occupations and industry sectors.

High-risk industry sectors (based on fatality rates) include agriculture, construction, food processing, and health care.

Workshop participants suggested several additional priorities for intervention criteria. Although data resources and workforce characteristics almost certainly vary by state it may be difficult to get state-by-state data. Typically Current Population Survey public datasets can be used only for assessing regional variations due to confidentiality requirements and small datasets in some states. Another priority for intervention is differences by country of origin in popular vocabulary for terms important for worker safety, such as names of equipment or tools, and cultural variations in perceptions of risk and need for protection. Development of strategies for moving toward a more common vocabulary related to safety and health is an important consideration, but the issue was not seen as the primary barrier to preventing injury and illness.

As described in the NIOSH National Occupational Research Agenda priorities related to organization of work, understanding the barriers and context within which public health interventions can be effective should be a priority in targeting Spanish-language materials, as it is in targeting English language public health interventions (NIOSH, 1996). In addition to language differences that can be addressed by translation, recognition of cultural differences may modify the nature of an effective public health intervention, even when addressing the same occupational health hazard in the same industry sector as an English-language counterpart. Effective interventions that extend beyond simple translation are important to address diversity within the Hispanic workforce and differences from the non-Hispanic workforce. Existing data to target, prioritize, and provide metrics for evaluation of the effectiveness of interventions is not currently limiting initial public health interventions but has considerable weaknesses. Workshop participants thought that major long-term initiatives should be developed by NIOSH, the Bureau of Labor Statistics, and OSHA to improve public datasets, and that several high-priority Hispanic target populations can be distinguished with relative confidence using current data.

3

What Information Exists and What Are the Gaps?

While it is clear from workshop discussions that occupational safety and health resources for Spanish-speaking workers in the United States are needed, it is less clear that there are adequate materials that fit the needs. Many different sources of Spanish-language materials were identified, primarily from the following domestic sources:

(1) governmental agencies (such as Occupational Safety and Health Administration [OSHA], Mine Safety and Health Administration, NIOSH, Environmental Protection Agency, and state programs);
(2) state governmental programs (especially in those states with a long history of participation of Hispanic immigrants in the labor force);
(3) labor unions who represent large numbers of Spanish-speaking workers;
(4) university and non-profit programs that serve immigrant workers;
(5) employer associations and insurance companies; and
(6) commercial material development firms that market their materials to employers of Spanish-speaking workers.

In addition, there are international sources of materials from the International Labor Organization, the Pan American Health Organization, and from many safety, health, and research institutions in Spanish-speaking countries.

A variety of problems were identified during the workshop regarding these existing materials.

(1) There is no centralized place to easily access existing resources. While there are some attempts to maintain collections of materials, these collections are not online, not reviewed, and not updated regularly. Even publicly funded materials in Spanish often are difficult to access (such as those developed under OSHA training grants).
(2) Only a small fraction of U.S. government materials has been produced in (or translated into) Spanish. It is unclear which criteria are applied to determine which materials to translate or whether there is an overall strategy to identify and address the need for materials in Spanish (and other languages).
(3) There is great variability in the quality of materials from all sources. For example, materials developed in other countries are often not relevant to a U.S. audience. Materials translated from English often suffer from poor translation, translation that is incorrect, or translation that is correct but overly technical and not easily understood by the intended audience.
(4) Frequently, those who are involved in training and education of workers find that the only materials available are highly technical documents, rather than practical teaching materials.
(5) There is much more material available for workers in agriculture and construction, while there is comparatively little for those who work in the service sector, food

processing, and small manufacturing, which have large numbers of monolingual Spanish speakers (see Appendix E).

(6) There is a particularly small amount of materials that addresses workers' health and safety legal rights, especially in a manner relevant to immigrant workers. For example, a recent Supreme Court decision has given the incorrect perception that OSHA laws do not apply to many immigrant workers (see Appendix E). There are few materials that address the intersection of immigration rights and health and safety rights.

At the workshop there was general agreement on the need to collect, evaluate and disseminate Spanish-language resources on an ongoing basis. This clearinghouse-type effort poses significant challenges, such as determining which criteria should be used to assess the quality of materials to be included. In order to do this it would be useful for NIOSH to call a consensus conference or workshop to focus specifically on this task. In addition, the workshop participants also agreed that identifying funding for active collection and maintenance of materials may prove difficult. However, given the paucity of effective educational materials for a large and growing U.S. population at risk, NIOSH, OSHA, and other agencies responsible for worker safety must rise to the challenge.

An additional point of agreement at the workshop was that it is not enough simply to collect, evaluate, and disseminate existing materials. Development of new materials would also be of great value. The need for these materials cannot be adequately met by a centralized governmental approach. This gap may best be filled by those who are in direct contact with the affected populations and aware of their health and safety training and information needs. Discussion during the workshop often centered on the important topic of providing resources for those who are best able to access high-risk Spanish-speaking workers and to support their efforts through:

(1) funding for tailored outreach and education efforts, and for development of appropriate materials needed for such efforts;
(2) networking among those who are reaching Spanish-speaking workers to exchange approaches and learn from one another's experiences;
(3) technical support for those community organizations that may have the best access to a particular worker population and may be the most trusted source of information and training for those workers, but may not have the necessary safety and health expertise to carry out the work without training and assistance.

4

Development and Evaluation of Materials

The first question regarding the development of new materials is how to prioritize which gaps to fill. The committee uses the term "materials" in this report to mean not only written information, but also communications methods such as audiocassettes, videocassettes, workshops, meetings, comic books, television and radio programming. In general, workshop participants were of the opinion that priority should not be on translation of technical health and safety documents. Rather, emphasis would be better placed on developing materials as part of a strategic initiative to reach Spanish-speaking workers, their employers, and their communities with practical information that can assist in injury and illness prevention in the workplace. Materials development in Spanish is important but only in the context of a larger outreach, education, and action strategy. A thought repeatedly expressed during the workshop was: "Good Spanish-language materials are necessary but not sufficient."

A more comprehensive approach requires that those who develop new, effective materials be able to plan a comprehensive education plan that takes into account the organizational, societal, economic, and other factors that define the environment in which workplace change must take place. Ideally, materials development would be in the context of a plan that considers the significant barriers immigrant workers may face in taking action for workplace safety. These include fear based on immigration status, linguistic and language barriers, and different work experiences in their countries of origin. An effective approach also requires that we draw on cultural strengths, such as strong community institutions and health beliefs.

Much is known about cultural appropriateness and cultural competence in other aspects of public health. In the opinion of the workshop participants, the occupational safety and health community needs to become more familiar with the literature and experiences in such fields as tobacco control, human immunodeficiency virus and acquired immune deficiency syndrome education, and other significant public health campaigns that have addressed cultural and linguistic issues successfully. Partnerships with community and other public health professionals with health education and communications expertise would be useful.

Finally, workshop participants noted that it would be beneficial if materials development in Spanish drew on sound principles of effective adult education, social marketing, and public health education. Some of the key precepts that are well established include:

(1) involvement of the target population in developing the educational message and outreach plan;
(2) pilot testing of materials with the target population to assess, for example, appropriate messages, literacy level and readability, terminology, and graphics;
(3) use of workers' own narratives and images in materials;
(4) development of action-oriented (problem-solving), participatory, and hands-on educational materials;
(5) addressing of different learning styles by using a variety of methods;

(6) use of multiple means of communication and education (e.g., radio, face-to-face, comic books); and
(7) design materials that allow them to be used with some flexibility in a variety of instructional and learning situations.

When existing materials are being translated into Spanish, many of these same principles will apply. Workshop participants agreed that materials should be adapted for use by a Spanish-speaking audience, rather than simply translated literally. There was substantial discussion in the workshop about the preference for materials developed originally in Spanish (or in Spanish and English) rather than translated. However, because this is not always feasible, the importance of high-quality translation was highlighted. In particular, translations should be performed by native Spanish speakers who are qualified as translators and not just by anyone who speaks Spanish; translations should then be reviewed by native Spanish speakers and pilot tested with the intended audience. When translating materials, the education level of the audience should be taken into account. Unfortunately, there are far too many examples of embarrassingly bad translations, according to the workshop participants.

Approaches to materials development are most effective if grounded in broader intervention efforts that are based on sound principles of program planning. These principles include targeting interventions based on the best available surveillance data; thoroughly assessing needs; involving stakeholders in the process; establishing clear, measurable objectives; designing a participatory education program; pilot testing with the target population; evaluating program effectiveness; and revising efforts based on findings. Workshop participants thought it is essential that leadership and staff in organizations that fund materials development and testing understand the resource needs. The CDC Guide to Community Based Public Health Practice includes methodological guidelines for designing and funding sound evidence-based intervention research.[3]

Once an educational intervention is well tested, a wide-ranging dissemination program will be needed. It is essential to customize dissemination program to each audience and situation. Given the seriousness of the situation, society cannot afford to continually "reinvent the wheel." Therefore, it is important that results be shared so that resources are maximized.

Workshop participants discussed the desire for a concerted effort to promote evaluation of Spanish-language materials. Evaluation should be built into the initial planning stages of materials development and educational interventions of all kinds. Materials development is greatly enhanced when target populations, objectives, and methods for measuring accomplishments are defined. However, during the push to develop and use materials, the evaluation phase does not always occur, because all too often these development projects lack evaluation resources.

Workshop participants thought that it is important to have a solid evaluation process of multiple phases, including (1) pilot testing and focus groups with intended audiences during the development phase; (2) feedback from end users (such as trainees) to determine comprehension of the materials; and (3) evaluation to determine which practices work best. In addition, there was discussion of how to review and rate existing materials in Spanish. One suggestion was that an organization should create a comprehensive clearinghouse of safety and health material on the Internet as a single source for this information in Spanish. Others responded that Internet sites might be burdensome and difficult to navigate, and the content quality might be questionable, depending on the skill and background of the person or group authoring a Web site. In addition to the task of developing acceptable criteria, the task of going through all the material may become overwhelming. As discussed in a later section, Hispanics still have limited access to the Internet.

There are, however, some positive examples of information compilation sites. The National Institute of Environmental Health Sciences clearinghouse on hazardous waste materials

---

[3] See <http://www.thecommunityguide.org>.

management[4] and the Building Trades Master Trainers forum[5] are sites with no official evaluation of materials. On the other hand, the National Agricultural Safety Database[6] and Electronic Library for Construction Occupational Safety and Health[7] have advisory boards to review content.

Workshop participants agreed that rating criteria are important. Possible rating criteria discussed include:

- whether materials for Hispanic workers are presented in a manner that is "culturally and linguistically competent," comfortable to the worker, and not offensive;
- whether materials are grammatically correct but in the vernacular of the workers, even when it may be a mix of Spanish and English (these variations can be identified as such);
- whether materials are created for the appropriate literacy levels of the workers;
- whether the information provided is technically sound; and
- whether the subject matter is compatible with audience priorities and needs.

Workshop participants agreed that it is essential that a solid materials evaluation process be implemented so that materials can be judged as efficacious before being introduced to the workforce. Likewise, a program of continual evaluation is important to assure that the target population understands the materials and the materials continue to improve. Agreement on a common set of evaluation criteria would be useful both for the development of new materials and for distribution of existing materials through some type of clearinghouse.

---

[4] See <http://www.niehs.nih.gov>. Date accessed November 5, 2002.
[5] See <http://www.buildingtrades.org>. Date accessed November 5, 2002.
[6] See <http://www.cdc.gov/niosh/nasd/ nasdhome.html>. Date accessed November 5, 2002.
[7] See <http://www.cdc.gov/niosh/elcosh>. Date accessed November 5, 2002.

5

Delivery Mode

Well-developed educational materials by themselves do not always assure that the workers who need them will be reached. As discussed earlier, it is important that materials be used in the context of a well-planed educational intervention. These interventions are best delivered by trusted sources and preferably in multiple ways. People are social in nature and adult learning principles tell us that the most effective learning occurs in small groups or units where there are personal relationships and an established level of trust. It is important that people are in a comfortable setting and trust those delivering the information. Frequently, safety and health practices are learned on the job site through peer training or in social settings where acquaintances discuss what they have learned. The importance of personal contact in these circumstances should not be underestimated.

There are many existing trusted channels of communication that could be used. For example, many local and state health departments, clinics, and community groups already have outreach for certain subjects. Incorporation of occupational safety and health information into current messages, courses, or information could make use of these networks.

Some community and grassroots organizations are providing workers with different services and may be able to share safety and health information at the same time. They may be able to implement the philosophy that giving people what they desire provides the opportunity to teach them something new. At the workshop an example was given of a group of women in a small village who wanted to learn to crochet. While it would have been difficult to gain their trust and teach them about health and hygiene from the outset, the instructor first taught them how to crochet. Then during the crochet exercises she taught them health and hygiene principles. In addition, the reason the women wanted to learn to crochet was to make items to sell in order to generate income to buy food for their children. Not only did the project gain their trust, it gave them additional means to improve the health of their families.

Many Hispanic people are religious and trust their local churches. Using the churches and their auxiliary groups is another way to reach Hispanic workers.

Schools present another important and largely unused channel of communication. Many Hispanic workers are migrant workers and move from area to area depending on the season. Sometimes their children are placed in schools with no bilingual programs or teachers. These children sit in class until it is time to move on and miss out on a learning opportunity. Children in school can be taught safety and health principles that they can pass on to their families or implement later in their lives. In order to do this, culturally and linguistically appropriate materials need to be given to children both for their own instruction, and to pass on to adults.

Within the limited-English, Spanish-speaking workforce, particularly among the recently arrived immigrants who work in high-risk fields, health- and safety-related information is exchanged orally and informally among co-workers. Effective strategies should consider ways to provoke discussion among workers. If such oral discourse can be encouraged, there is improved retention of information (both by the worker who communicates the information and the listener). There has been a good deal of attention in Latin America to these dynamics in connection with "popular education" campaigns and in the United States as part of efforts such as HIV prevention campaigns.

Another mode of communication is the use of new technology, such as computer-based learning and access to information on the Internet. However, these channels pose a dilemma. Many people in our society look to the Internet for information and education, and it can contain an enormous amount of information helpful to workers. The Internet can be a great forum for sharing and exchanging information. Internet forums such as listserves, clearinghouses, and chat rooms can provide up-to-date training or informational materials for people who know how to use them. However, it contains so much information that it may be difficult to find appropriate and well-developed materials. Access to the Internet is increasing rapidly. However, there is a significant "digital divide"—only 31.6 percent of Hispanics use this tool (U.S. Department of Commerce, 2002). If the target audience is Hispanic workers with low literacy levels, the percentage shrinks even further. Even when the Internet is accessed it lacks the human contact and group interaction element that is essential for effective adult education. On balance it was felt that the Internet is a better tool for disseminating material to trainers who in turn will teach Hispanic workers, rather than as a primary tool for reaching Spanish-speaking workers directly.

One important mode of dissemination that came up frequently during the workshop is the Spanish-language mass media. There are Spanish newspapers throughout the United States, and many metropolitan areas may have several of these newspapers. Many rural areas with concentrations of Latino immigrants have weekly Spanish newspapers. There has also been an increase in Spanish-language magazines. While periodicals reach many people, there is still the obstacle of low literacy levels among workers. Fortunately the Spanish media are not limited to the written word, and the number of Spanish television and radio stations is rapidly increasing. In the past, information from television and the radio was limited to public service announcements. There are currently a variety of options available, such as providing subjects for talk shows, news shows, and integrating health and safety into regular television programming. Another advantage of mass media is that the media can reach multiple important audiences, including workers, small employers, and even families of workers, all of whom may have no other way to receive the information. Again, there is the issue of program effectiveness seldom being systematically evaluated. As recommended in a recent Institute of Medicine report, it is essential to begin funding and conducting appropriate evaluations of such large, costly, and potentially important interventions (NRC, 2002).

Because many Hispanic workers come from Mexico, it was suggested that collaboration with Mexican agencies could benefit both countries by teaching workers safety and health while they are in Mexico. Therefore the workers would be better prepared if they were to come to the United States. There are many other possible channels of dissemination, and they include:

(1) unions representing large numbers of Spanish-speaking workers;
(2) Spanish-speaking and other employers;
(3) universities and colleges (United Association for Labor Education);
(4) National Injured Workers Group;
(5) Mexican consulates;
(6) trade associations;
(7) suppliers (e.g., personal protective equipment, chemicals);
(8) immigrant rights groups;
(9) small business centers (Small Business Administration);
(10) U.S. government agencies (OSHA, NIOSH);
(11) other government agencies, such as Mexican consulates and the Mexican Secretaria de Relaciones Exteriores Instituto para Mexicanos en el Exterior, and Secretaria de Salud;
(12) workers' centers; and
(13) adult education classes.

It is clear that there are many modes, methods, and channels of disseminating information. Some have proven to be successful in areas where the Hispanic community has had time to become established. However, Hispanic workers are now moving to diverse areas throughout the entire United States. In order to reach these workers with safety and health training and information it may be best to implement a combination of strategies.

6

Overarching Issues

This section summarizes the material presented in the earlier chapters and discusses several overarching issues that came up repeatedly during the workshop.

At the workshop, discussion often centered on the underreporting of incidents and severity and the non-reporting of violations. Spanish-speaking workers often are less aware of regulations and are less likely to report violations than other workers, as some may be concerned about deportation (and sometimes are threatened with this). Hispanics are often aware that they are easily replaceable, so are less likely to report violations and are reluctant to miss work because they might lose their pay or their job. Spanish-speaking workers may be less likely to use a company health service department if their injuries or other harms are noted and reported.

Occupational safety and health information in Spanish is necessary but not currently sufficient to resolve these issues. Workshop participants agreed that communication materials and strategies must be in Spanish, but that language alone is not enough for adequate communication. It is essential that messages be delivered in culturally appropriate ways, both in terms of content and approaches (see Appendix E and Chapter 3 of this report for details). Workshop participants also agreed that the quality of existing material is mixed, with some of it very poor. To address the issue of quality of material, workshop participants urged that evaluation standards for materials be implemented, including an evaluation of existing materials. They suggested that a national clearinghouse (ideally, on the Internet) of good-quality materials be established with information on best practices.

Workshop participants also identified best practices with respect to the content of occupational safety and health material in Spanish. The language must be appropriate to the educational level of the target audience. When several Spanish terms exist, it may be necessary to include more than one. Messages must be simple, brief, and clear. It may be necessary to use *spanglish*[8] if that is what workers use (e.g., *sheetrockero*, for a person who works with sheetrock). It may be important to communicate some basic safety English (e.g., "duck" and "run"). It is important to use participant observation at worksites to develop materials, and to understand barriers and incentives tied to workers' beliefs and experiences. Participant observation is a tool developed by anthropology that involves the researcher spending time in the community (or worksite) alongside the workers. It allows the researcher or evaluator to experience the conditions, difficulties, facilitators, language, and so on at the worksite (Scrimshaw and Hurtado, 1987). The material should pinpoint behaviors that are key to safety and focus on these. It is essential to develop both content and delivery modes for messages with worker input.

The workshop participants agreed that multiple modes of delivery should be considered, not just print media. Preferred delivery modes include *photonovelas*,[9] radio, TV (talk shows as well as ads and announcements), websites, community sites (e.g., schools, churches), and

---

[8]A mixture of Spanish and English in words or sentences.

[9]*Photonovelas* are essentially expanded comic books, or simple, short highly illustrated dramatic stories. For example, a protagonist in the story might get injured on the job, and readers would learn how injuries occur and how to prevent them.

worksite training sessions. Again, it is essential to develop these with worker input. In several areas, workshop participants had differing views of how to proceed. There was much discussion on the issue of whether to use a clearinghouse or local groups for development of materials. Some participants thought that a national clearinghouse, which developed and recommended materials, is most effective. Others thought that local development by those close to the culture and the worksite would result in more appropriate and innovative approaches. A combination of local and national approaches might make the most sense.

There was also considerable discussion and varied opinions about the best approach to translating material into Spanish. There was agreement that direct translation of materials, including warning labels and signs, usually does not convey the correct meaning, and that adaptation is essential at the very least. Adaptation involves working with the materials to convey the correct meaning in culturally, educationally, and linguistically acceptable language. Some workshop participants thought that adapting would carry too much English or U.S. cultural and linguistic "baggage," and that it was essential to create materials in Spanish, with user participation.

Workshop participants debated the extent to which various government agencies involved in worker safety had worked together effectively in the past. There was agreement that working together is important, and that the dialogue during the workshop between representatives of various agencies and with other workshop participants was useful in furthering such collaborations. It was noted that partnerships of industry, OSHA, NIOSH, and worker groups provide models for successful training programs.

Audiences for improved worker safety communication materials and for the methods for developing and improving these include:

(1) owners and employers;
(2) controlling contractors;
(3) frontline supervisors;
(4) trainers, safety personnel;
(5) workers, including day laborers (There is great variety in type of work, risks, level of training, literacy.);
(6) families;
(7) unions, trade associations;
(8) federal agencies (e.g., NIOSH, OSHA, Environmental Protection Agency);
(9) local and state health departments and clinics;
(10) community organizations;
(11) insurers, bonding agencies, financial organizations;
(12) media; and
(13) policy makers (local, state, national, international).

Safety for workers involves behavior changes to ensure safe habits and safe conditions. The workshop participants discussed the need to change employer, supervisor, and worker behaviors. It was noted that workers seldom have the power to change working conditions. This power lies with owners, employers, contractors, supervisors, etc., and communication with these groups regarding their responsibilities for worker safety is essential. Intermediaries (e.g., plant supervisors, contractors) can play an important role ensuring safety and delivering informational materials. Often, the solution to high-risk situations lies in changing the environment rather than, or more than, changing the worker behavior.

Workshop participants discussed in detail the role of the government at federal, state, and local levels and agreed that key roles include:

(1) setting goals and priorities for worker safety (e.g., Healthy People 2010, surgeon general's reports);
(2) establishment of regulations;
(3) monitoring and enforcement of compliance with regulations;
(4) considering goals in resource allocation for research, surveillance, enforcement, and materials development;
(5) development and communication of essential content of worker safety notices and information;
(6) remedying the lack of legal requirement to communicate with workers in their own language (states are not consistent on this, nor are industries); and
(7) clearinghouse function for materials and rating of materials.

The range of agencies involved is broader than many would think (e.g., Health and Human Services, NIOSH, National Institutes of Health, CDC, Environmental Protection Agency, Department of Transportation, Food and Drug Administration, Department of Labor, OSHA, U.S. Department of Agriculture);

Workshop participants also discussed research needs for developing occupational health and safety information in Spanish. There was consensus that the top two research needs were funding to develop and test materials (intervention studies) at local universities, community organizations, and others and funding for better documentation of risky behaviors and exposures for Spanish-speaking workers.

There was also a discussion regarding the need for a multi-national approach. Workshop. participants thought that the United States could work with representatives of Spanish-speaking countries to assess, develop, and improve materials in the context of home countries. The United States could also work with Mexico in particular (and with Mexico's Secretary for Health) to develop materials for use with workers in both countries, and for training workers who will immigrate or who move back and forth across the border. It may be useful for the United States to work with the U.S.-Mexico Border Health Association, the U.S.-Mexico Foundation for Science, the Pan American Health Organization and similar organizations. And finally the United States should work with *maquiladora* health and safety support networks (American Public Health Association, American Industrial Hygiene Association, Mexican Industrial Hygiene Association).

The workshop participants were greatly appreciative of the opportunity provided by NIOSH for them to come together to discuss these issues. The diversity of the participants was noted. Participants included representatives from government agencies, community organizations, academic research centers, employers, outreach workers, and union members. Discussions were open, honest, and productive. Together there was agreement on the importance of the problem, particularly in relation to the numbers of workers affected, the risks inherent both in common occupations for Latinos and in workers with little or no command of English, and in the lack of power to affect change or ask for their rights. The power differential between workers and employers is particularly great because of the lack of legal status of many workers, even when employers recruit across the border. The need to protect all workers, as reflected in the NIOSH initiative in convening this conference, was praised by the participants.

References

Center to Protect Workers' Rights (CPWR). 2002. *The Construction Chart Book*, Third Edition. Silver Spring, Md., <http://www.cpwr.com> Page 33, date accessed: November 5, 2002.

CDC (Centers for Disease Control and Prevention). 1993. Use of race and ethnicity in public health surveillance. Summary of the CDC/ATSDR Workshop, MMWR 42(No.RR-10).

Hahn, R. 1992. The state of federal health statistics on racial and ethnic groups. *Journal of the American Medical Association* 267(2):268-71.

Hahn, R. 1999. Why race is differentially classified on U.S. birth and infant death certificates: An examination of hypotheses. *Epidemiology* 19(2):108-111.

Hahn, R., and D. F. Stroup. 1994. Race and ethnicity in public health surveillance: Criteria for the scientific use of social categories. *Public Health Reports* 109(1):7-15.

Hiles, D. R. 2001. A first look at employment and wages using North American Industrial Classification System. *Monthly Labor Review* December, p. 22-31.

McKay, R. 1992. Bureau of Labor Statistics "Cultural Factors Affecting Within Household Coverage and Proxy Reporting in Hispanic (Salvadoran) Households. A Pilot Study," Proceedings of the Section on Survey Research Methods, American Statistical Association, p. 614-618.

NIOSH (National Institute for Occupational Safety and Health). 1996. National Occupational Research Agenda. Available online at <http://www.cdc.gov/niosh/nora.html>, date accessed: November 5, 2002.

NRC (National Research Council). 2000. Promoting Health: Intervention Strategies from Social and Behavioral Research. Washington, D.C.: National Academy Press.

NRC. 2002. Speaking of Health: Assessing Health Communication Strategies for Diverse Populations, Washington, D.C.: National Academy Press.

Rodriguez, G. 1999. *From Newcomers to New Americans: The Successful Integration of Immigrants into American Society*. Washington, D.C.: National Immigration Forum.

Scrimshaw, S. C. M., and E. Hurtado. 1987. Rapid Assessment Procedures for Nutrition and Primary Health: Anthropological Approaches to Programme Improvement. Published jointly by the UCLA Latin American Center (Los Angeles), the United Nations University (Tokyo), and UNICEF. Spanish edition, 1988; French edition, 1991; Chinese edition, 1991.

Therrien, M., and R. R. Ramirez. 2000. *The Hispanic Population in the United States: March 2000,* Current Population Reports, p. 20-535. Washington D.C.: U.S. Census Bureau.

U.S. Census Bureau. 2001a. The Hispanic Population of the United States, March, 2000. Washington, D.C.: U.S. Census Bureau.

U.S. Census Bureau. 2001b. Population by Race and Hispanic or Latino Origin for the United States. 1990 and 2000 U.S. Bureau of Census April 2, 2001. Census PHC-T-1. Washington, D.C.: U.S. Census Bureau.

U.S. Department of Commerce. 2002. A Nation Online: How Americans are Expanding Their Use of the Internet. Washington D.C. pp. 106.

Walker J. A., and J. B. Murphy. 2001. Implementing the North American Industry Classification System at Bureau of Labor Statistics. *Monthly Labor Review* December:15-21.

APPENDIXES

Appendix A

Biographical Sketches of Committee Members

SUSAN C. SCRIMSHAW, *chair,* has been the dean of the School of Public Health at the University of Illinois at Chicago since December 1994. Her major interests are medical anthropology, human reproduction, demography, applied anthropology, violence prevention, and culture change with a focus on Latin America, the Caribbean, and U.S. Latinos. She is a member of the Institute of Medicine and has served on various NRC committees. She is currently a member of the Board on Global Health and the chair of the Committee for Behavior Change in the 21st Century: Improving the Health of Diverse Populations. Dr. Scrimshaw received her Ph.D. from Columbia University, Department of Anthropology, with a focus on Latin America, medical anthropology, and demographics.

ROBIN BAKER is director of the Labor Occupational Health program at the Center for Occupational and Environmental Health in the School of Public Health at University of California, Berkeley. She administers programs that provide training, information, and technical assistance for labor, management, and health professionals. Her role includes the development of written materials, teaching of courses, planning conferences, and conducting research. Prior to that she was project director for the Santa Clara Center for Occupational Safety and Health (Project on Health and Safety in Electronics). Ms. Baker received her M.P.H. in health education from the University of California, Berkeley.

LAWRENCE SHAW-SALAZAR is an industrial hygienist at IHI Environmental. He specializes in industrial hygiene monitoring, management of asbestos projects, indoor air quality investigations, and facility health and safety inspections. He previously served as chief consultant and owner of LRS Spanish Safety Consultants, where he developed and taught Spanish safety- and health-training classes; translated safety documents into Spanish; performed industrial hygiene sampling; and performed site safety and health compliance audits. He was also a health and safety consultant for Phelps Dodge Mining Company, where he developed a corporate safety and health manual and translated policies and training materials into Spanish. He received his B.S. and M.S. in environmental health, and industrial hygiene from California State University, Fresno, and University of Washington, Seattle, and his J.D. from Brigham Young University.

JAMES PLATNER is associate director at the Center to Protect Workers' Rights (CPWR), a research and training institute of the Building and Construction Trades Department, AFL-CIO. Dr. Platner co-authored a working paper on Hispanic occupational injuries and fatalities in construction. CPWR has developed Spanish-language safety and health materials and instructions for use in training residential construction workers and for hazardous waste operator training, asbestos abatement, and lead abatement worker and supervisor training. He has a B.S. in biophysics from Johns Hopkins University and an M.S. in radiation biology and a Ph.D. in toxicology from the University of Rochester School of Medicine.

NRC Staff

TAMARA L. DICKINSON, *Study Director,* is a senior program officer with the NRC's Board on Earth Sciences and Resources, responsible for managing the activities of the Committee on Earth Resources. Dr. Dickinson has been awarded the National Academies individual distinguished service award. She has served as program director for the Petrology and Geochemistry program in the Division of Earth Sciences at the National Science Foundation. She has also served as discipline scientist for the Planetary Materials and Geochemistry program at NASA headquarters. As a postdoctoral fellow at NASA Johnson Space Center she conducted experiments on the origin and evolution of lunar rocks and highly reduced igneous meteorites. She holds a Ph.D. and an M.S. in geology from the University of New Mexico and a B.A. in geology from the University of Northern Iowa.

KAREN L. IMHOF is a senior project assistant for the NRC's Board on Earth Sciences and Resources. She previously worked on the Board on Agriculture and Natural Resources. Before coming to the National Academies she worked as a staff and administrative assistant in diverse organizations, including the National Wildlife Federation and the Three Mile Island nuclear facility.

TANJA PILZAK is a research assistant with the Board on Agriculture and Natural Resources (BANR). She previously served the National Academies in the Office of Contracts and Grants, first as a contract assistant and then as a proposal specialist. Prior to coming to the National Academies she worked with students in the Department of Resident Life at the University of Maryland, College Park.

Appendix B

Workshop Agenda

WORKSHOP ON COMMUNICATING OCCUPATIONAL SAFETY AND HEALTH INFORMATION TO SPANISH-SPEAKING WORKERS

National Research Council
Manchester Grand Hyatt
San Diego, CA
May 29-30, 2002

Wednesday, May 29

Windsor Room, Manchester Grand Hyatt, San Diego, CA

8:00 A.M.	Continental breakfast available in meeting room	
8:30 A.M.	Introductions and review of agenda	*Susan Scrimshaw* *Committee Chair*
8:45 A.M.	Demographics and distribution in the U.S. workforce of Spanish-speaking workers and employers	*Scott Richardson* *Bureau of Labor Statistics*
9:00 A.M.	Discussion and summary	
10:30 A.M.	Break	
10:45 A.M.	Discussion and summary (continued)	
12:00 P.M.	Lunch	
1:15 P.M.	Examination of occupational safety and health materials currently available in Spanish	*Marianne Parker Brown* *UCLA-LOSH*
1:30 P.M.	Discussion and summary	
3:00 P.M.	Break	
3:15 P.M.	Discussion and summary (continued)	
4:30 P.M.	Adjourn	

Thursday, May 30

Windsor Room, Manchester Grand Hyatt, San Diego, CA

8:00 A.M.	Continental breakfast available in meeting room	
8:30 A.M.	Examination of the informational needs of Spanish-speaking workers, their employers and occupational safety and health practitioners	*Tom O'Connor* *Consultant, NC Committee on Occupational Safety and Health*
8:45 A.M.	Examination of the occupational risks and occupational safety and health communication needs of Spanish-speaking children who are employed or live on farms	*Martha Vela Acosta* *University of Texas, Brownsville*
9:00 A.M.	Examination of the occupational risk and occupational safety and health communication needs of Spanish-speaking workers employed in mining, construction and manufacturing, with emphasis on meatpacking and assembly	*Raphael Moure-Eraso* *University of Massachusetts, Lowell*
9:15 A.M.	Breakout groups	
10:30 A.M.	Break	
10:45 A.M.	Breakout groups (continued)	
12:00 P.M.	Lunch	
1:15 P.M.	Final discussion and summary session to discuss overarching issues	
3:30 P.M.	Discussion with National Institute for Occupational Safety and Health representatives	
4:00 P.M.	Adjourn	
6:30 P.M.	Committee dinner	

Friday, May 31

Windsor Room, Manchester Grand Hyatt, San Diego, CA

Closed Session

7:30 A.M.	Until noon

Appendix C

Workshop Participants

Alex Alvarez
Communications Specialist
Centers for Disease Control (CDC) en Espanol
1600 Clifton Road
Atlanta, GA 30333

Robin Baker
Director, Labor Occupational Health Program
Center for Occupational and Environmental Health
University of California, Berkeley
2223 Fulton Street, 4th Floor
Berkeley, CA 94720-5120

Sherry Baron
Medical Officer
Division of Surveillance, Hazard Evaluation, and Field Studies
NIOSH
4676 Columbia Road, MS R10
Cincinnati, OH 45226-1998

Deanna Berrett
Director, Marketing Occupational Safety and Health
National Safety Council
Itasca
1121 Spring Lake Drive
Chicago, IL 60143

Todd Briggs
Program Specialist
National Safety Council
1025 Connecticut Avenue, N.W.
Suite 1200
Washington, DC 20036

Marianne Parker Brown
Director, LOSH Program
University of California at Losa Angels
Hershey Hall
Los Angeles, CA 90095-1478

Felipe Devora
Director of Safety Services
Fretz Construction Company
6301 Long Drive
Houston, TX 778097

Susan E. Feldmann
Writer Editor
Education and Information Division
NIOSH
4676 Columbia Parkway, MS C18
Cincinnati, OH 45226-1998

Sarah Felknor
Assistant Professor
School of Public Health
University of Texas
P.O. Box 20186
Houston, TX 77225-0186

Beatriz Gonzalez
AmeriCorps Pesticide Safety Trainer
Association of Farmworker Opportunity Programs - SAFE Program
1900 N. Dinuba Blvd.
Visalia, CA 93291

Ben Hart
Manager, Mine Safety & Health Training Program
Florida Department of Environmental Protection
2051 E. Dirac Drive
Tallahassee, FL 32310

Theresa Laing
Worker Trainer
PACE International Union
P.O. Box 115
Crossett, AR 71365

Herbert Linn
Acitivity Chief, Information Management and Dissemination Activity
NIOSH/DSR/IMDA
1095 Willowdale Road
Morgantown, WV 26505-2888

Max Lum
Associate Director of Health Communication
NIOSH
200 Independence Avenue, S.W.
Washington, DC 20201

Charlene Maloney
Publications Management Specialist
Education and Information Division
NIOSH
4676 Columbia Parkway, MS C18
Cincinnati, OH 45226-1998

Elaine Mann
Technical Information Assistant
Education and Information Division
NIOSH
4676 Columbia Parkway, MS C19
Cincinnati, OH 45226-1998

Fernando Marroquin
Biomedical Services Manager
University of Alabama
Box 870388
Tuscaloosa, AL 35487-0388

Rafael Moure-Eraso
Professor, Department of Work Environment
University of Massachusetts, Lowell
One University Avenue
Kitson 200
Lowell, MA 01854-2867

Jackie Nowell
Director, Occupational Safety and Health Office
UFCW International Union
1775 K Street, N.W.
Washington, DC 20006

Tom O'Connor
Consultant
North Carolina Committee on Occupational Safety and Health
133 Circadian Way
Chapel Hill, NC 27516

Andrea Okun
Deputy Director, Education and Information Division
NIOSH
4676 Columbia Parkway, MS C14
Cincinnati, OH 45226-11998

Clara Olaya
Health Communication Specialist
Centers for Disease Control
1600 Clifton Road
Altlanta, GA 30333

Elaine M. Papp
Program Analyst
OSHA
200 Constitution Avenue, NW
Washington, DC 20210

Yolanda Partida
Director, National Program Office
Hablamos Juntos
1051 N. Mills Avenue
Claremeont, CA 91711

Robert H. Peters
Supervisory Social Scientist
Mining Injury Prevention Branch
Pittsburgh Researc Lab
CDC
NIOSH
Cochrans Mill Road
Pittsburgh, PA 15236

James Platner
Associate Director
Center to Protect Workers Rights
8484 Georgia Avenue
Suite 1000
Silver Spring, MD 20910

Eddie Ramirez
AmeriCorps Pesticide Safety Trainer
Association of Farmworker Opportunity Programs, SAFE Program
P.O. Box 128
Five Points, CA 93624

Joe Reina
Deputy Regional Administrator
Department of Labor/OSHA
525 S. Griffin Street Street
Room 602
Dallas, TX 75202

Scott Richardson
Program Manager, CFOI
Bureau of Labor Statistics
Postal Square Building
2 Massachusetts Avenue, N.E.
Washington, DC 20212-0001

Richard Rodriguez
Assistant Deputy Director
California Department of Health Services
714 P Street, Room 1492
Sacramento, CA 95814

Elsa Roman
Health and Safety Specialist (Training)
Mine Safety and Health Administration, EFS West
300 San Mates Blvd., Suite 407
Albuquerque, NM 87108

Susan Scrimshaw
Dean, School of Public Health
University of Illinois at Chicago
1603 West Taylor Street, MC 923
Chicago, IL 60612

Larry Shaw-Salazar
Industrial Hygienist
IHI Environmental
4527 North 16th Street
Suite 105
Phoenix, AZ 85016

Rosemary Sokas
Associate Director for Science
NIOSH/CDC
Room 715 H
200 Independence Avenue, S.W.
Washington, DC 20201

Andrea Steege
Epidemiologist
Divison of Surveillance, Hazard Evaluations, and Field Studies
NIOSH
4675 Columbia Parkway, MS R18
Cincinnati, OH 45226

Doug Stephens
Coordinator of Grant H&S Field Operations
PACE International Union
P.O. Box 1475
Nashville, TN 37202

Marie Haring Sweeney
Chief, Document Development Branch
NIOSH
Education and Information Division
Mailstop C32, 4676 Columbia Parkway
Cincinnati, OH 45226

Suzanne Teran
CA WISH Coalition
Labor Occupational Health Program
University of California at Berkeley
2223 Fulton St.
Berkeley, CA 94720-5120

Martha Soledad Vela-Acosta
Associate Professor
University of Texas, School of Public Health
Health Science Center
RAHC Building 1220D, 80 Fort Brown
Brownsville, TX 78520

Tammy Dickinson
Sr. Program Officer
National Research Council

Tanja Pilzak
Research Assistant
National Research Council

Karen Imhof
Sr. Project Assistant
National Research Council

Appendix D

Hispanic Workers in the United States: An Analysis of Employment Distributions, Fatal Occupational Injuries, and Non-fatal Occupational Injuries and Illnesses

Scott Richardson, Bureau of Labor Statistics
John Ruser, Bureau of Labor Statistics
Peggy Suarez, Bureau of Labor Statistics

INTRODUCTION

The United States Census Bureau predicts that by the year 2050 Hispanics will represent one out of every four persons in the United States, up from about one in eight in 2000. Driven largely by immigration, this dramatic growth in the Hispanic population will continue to present new challenges in health care, education, and the workplace.

The results from the 2000 Census of Fatal Occupational Injuries (CFOI) program show higher fatal work injury rates for Hispanic workers than for other racial/ethnic groups--rates that appear to be increasing even as fatal work injury rates for most other United States workers are declining. Non-fatal occupational injury and illness rates are also higher among Hispanic workers.

Employment distributions tell us that Hispanic workers tend to be more heavily represented in higher-risk industries and occupations than non-Hispanic whites and other racial/ethnic groups. The question becomes to what extent are these higher fatality rates explained by this disproportionate representation of Hispanics in higher-risk industries and occupations. Also, what differences can be seen between the experience of foreign-born Hispanic workers and native-born workers in the occupational injury and illness data?

National data also tell us that the challenges in occupational safety and health are not limited to those states traditionally associated with large Hispanic populations, such as California or Texas but will impact numerous other states not traditionally known for large Hispanic populations. Hispanic communities are growing rapidly in states such as North Carolina, Arkansas, and Georgia. Moreover, the nature of the occupational injuries and illnesses differs from State-to-State and is largely determined by the industries within each state.

It is necessary to have reliable and comprehensive data to formulate appropriate and measurable strategies to address these challenges. In terms of surveillance of occupational injuries and illnesses for Hispanic workers, the data tell an interesting but incomplete story. Many gaps still exist in the data, especially in the non-fatal injury and illness data for Hispanic workers.

This paper summarizes the data on the demographics of the Hispanic population (including employment data) and presents an overview of the available surveillance data for fatal work injuries and non-fatal occupational injuries and illnesses for Hispanic workers. The data are from the Bureau of Labor Statistics' Occupational Safety and Health Statistics (OSHS) programs.

HISPANICS IN THE UNITED STATES: A DEMOGRAPHIC AND SOCIOECONOMIC OVERVIEW

Growth of the Hispanic Population in the United States

In 2000 there were 35.3 million Hispanics in the United States, a 58 percent increase over the 22.4 million Hispanics recorded by the Census Bureau in 1990 (Census Bureau 2001 [1], Census Bureau 2001[2]). This increase follows an increase of 53 percent in the United States Hispanic population between 1980 and 1990 (National Council of La Raza 2001). In 1990 Hispanics represented about 9 percent of the population. By 2000 the representation of Hispanics had grown to 12.5 percent (Census Bureau 2001 [1]). In 2050 it is predicted that Hispanics will represent one out of every four persons in the United States, up from about one in eight today (Census Bureau 2001 [3]).

While United States Hispanics of Mexican origin make up the largest segment of the country's Hispanic population, those with origins in other Spanish-speaking countries and regions are also strongly represented among United States Hispanics (Census Bureau 2001 [1],) Table A presents the distribution of United States Hispanics by country of origin.

TABLE A Distribution of U.S. Hispanics by Type, from Census 2000

	U.S. Hispanic Population	Total U.S. Population
Total Number	35.3 million	281.4 million
Percentage	100%	100%
Mexico	58.50%	7.30%
Puerto Rico	9.60%	1.20%
Cuba	3.50%	---
Other Hispanic	28.40%	3.50%
Dominican	2.20%	---
Central America	4.80%	---
Salvadoran	1.90%	
Guatemalan	1.10%	
Honduran	0.60%	
South America	3.80%	---
Colombian	1.30%	
Ecuadorian	0.70%	
Peruvian	0.70%	
Spaniard	0.3%	---
Other	17.30%	2.10%

SOURCE: U.S. Census Bureau

Growth of Hispanic Population by State

When looking at increases in the Hispanic population by state over the last decade (including both immigrant and non-immigrant Hispanics), it is clear that the states with traditionally large Hispanic populations continued to show the greatest numerical growth.

California and Texas continue to have the largest Hispanic populations, and about half of all United States Hispanics live in those two states (Census Bureau 2001 [4]).

Table B presents the 10 states that had the largest Hispanic populations in 2000. These states are also among the states recording the sharpest increases in the number of Hispanics between 1990 and 2000. California, for example, added over 3.2 million to its Hispanic population during that time. Texas added 2.3 million and Florida's Hispanic population increased by 1.1 million (Census Bureau 2001 [4], Census Bureau 2001 [5]).

In looking at the growth of Hispanic population in terms of percent change since 1990, evidence of strong growth in states not traditionally associated with large Hispanic populations emerges. Table C presents a list of the 10 states with the largest percent change in population since 1990 (Census Bureau, 2000).

TABLE B States with Largest Hispanic Populations, 2000

State	Total Hispanic Population, 2000 (in millions)
California	11.0
Texas	6.7
New York	2.9
Florida	2.7
Illinois	1.5
Arizona	1.3
New Jersey	1.1
New Mexico	0.8
Colorado	0.7
Washington	0.4

SOURCE: U.S. Census Bureau

TABLE C Top 10 States by Hispanic Percent Change Since 1990

State	Percent Change
North Carolina	394
Arkansas	337
Georgia	300
Tennessee	278
Nevada	217
South Carolina	211
Alabama	208
Kentucky	173
Minnesota	166
Nebraska	155

SOURCE: U.S. Census Bureau

Growth of Hispanics by Metropolitan Area

Hispanics live largely in urban areas. More than 9 in 10 Hispanics live within a metropolitan area and nearly half of all Hispanics live in a central city within a metropolitan area.

Only 9 percent of Hispanics live outside metropolitan areas, as compared to nearly one-fourth of non-Hispanic whites (National Council of La Raza 2001).

In 2000 the largest Hispanic populations were in New York (2.2 million), Los Angeles (1.7 million), Chicago (.8 million), Houston (.7 million), and San Antonio (.7 million) (Census Bureau 2001 [4], Census Bureau 2001 [5]). However, the metropolitan areas with the largest percentage increases between 1990 and 2000 were again ones that may not be traditionally associated with large Hispanic populations. The top three fastest growing Hispanic communities are in North Carolina. The five metropolitan areas with the largest growth in their Hispanic populations were Greensboro (694 percent growth), Charlotte (622 percent), Raleigh (569 percent), Atlanta (362 percent), and Las Vegas (262 percent).

Immigration

Much of the increase in the Hispanic population in the U. S. is driven by immigration. According to United States Census Bureau data, about half of the foreign-born population in the United States is from Latin America.

Since 1970, 8 of the top 20 countries of birth for immigrants are Spanish-speaking countries. Table D presents those eight countries along with the number and percent of immigrants from those countries residing in the United States (Camarota 2001). In 2000 about two out of every five Hispanics in the United States was born in a country other than the United States (Census Bureau 2000, Census Bureau 2001 [2], Census Bureau 2001 [3], Census Bureau 2001 [4], Census Bureau 2001 [5]).

TABLE D Top Spanish-speaking Immigrant Countries of Birth

	Number of Immigrants (000s)	Percentage of Total Immigrants
Total (all immigrants)	28,379	100
Mexico	7,858	28
Cuba	952	3
El Salvador	765	3
Dominican Republic	692	2
Columbia	435	2
Peru	328	1
Guatemala	327	1
Ecuador	281	1

NOTES:
1. Totals may include categories not shown separately.
2. Percentages may not sum to total due to rounding.

SOURCE: U.S. Census Bureau

In general, immigrants tend to have lower educational attainment, greater poverty, and less income than the native, non-immigrant population. Of the immigrants who arrived during the 1990-1999 period more than a third (34.4 percent) have less than a high school education. This percentage has risen for each of the past four decades, up from about one-fifth (19.3 percent) of immigrants who arrived in the years prior to 1970. The poverty rate for immigrants is 50 percent higher than the rate for the native-born. Median income for immigrant workers is $23,000 or nearly 25 percent lower than the median income for the native born population. The Center for Immigration Studies estimates that over a third of all unskilled jobs are now held by immigrants (Camarota 2001).

With regard to the number of undocumented workers, estimates vary. The Immigration and Naturalization Service (INS) estimates that the number of new undocumented aliens joining the United States population each year is about 420,000. After adjustments are made for deaths, emigration, and changes in legal status, this flow of undocumented workers decreases to a net of 275,000 annually (Camarota 1997). With the changes in border security imposed after the September 11, 2001, events, this number is expected to decrease.

NAFTA has altered the trends in illegal immigration. A larger percentage of migrating workers are opting to stay on the Mexican side of the border to live and work. The population on the Mexican side of the border has increased by nearly 50 percent over the last decade, while the United States side of the border showed an increase of only about one-fifth over that same period. (Economist 2001). This trend is likely to continue given the tightened border security instituted after the events of September 11.

RESULTS

Employment

Data from the United States Current Population Survey indicate that 13.8 million Hispanics were employed in the civilian workforce on average each year between 1998 and 2000 (see Table E1), as compared to 14.5 million black non-Hispanics and 99.1 million white non-Hispanics. Nationwide, 11.5 percent of male civilian employees and 9.1 percent of female civilian employees were Hispanic.

The service industries and retail trade employed the most Hispanic men and women, with construction industries placing third in employment among Hispanic men. Service industries and retail trade are also the industries that employ the most white and black men and women. Agriculture, forestry, and fishing had the highest percentage of male workers who were Hispanic (25.3), while both white and black men were under-represented in this industry. Hispanic men made up 14.5 percent of construction employment. In contrast, the proportion of black men in construction (6.3 percent) is below their overall percentage of employment (9.5 percent), while the percentage of white men in construction is slightly higher than their percentage in all jobs. Hispanic men also make up a high percentage of the very small private household industry.

Hispanic women accounted for 29.4 percent of female private household workers, a percentage that is far higher than their overall percentage of female employment. In comparison, a slightly higher percentage of black women work in private household jobs (14.8 percent) than in all jobs (12.5 percent), while white women are substantially under-represented in this industry division. Hispanic women are also well represented in nondurable goods industries (14.8 percent), with black women in this industry making up a slightly higher percentage (13.2) than they do in all industries. In comparison, black women make up especially high percentages of female employment in transportation, public utilities, and public administration.

Two states accounted for just over one-half of the total Hispanic employment: California with 30.9 percent of the national total, or 4.3 million workers, and Texas with 20.2 percent of the national total, or 2.8 million workers (see Table E2). Florida and New York were third and fourth in terms of the number of employed Hispanics, with 1.3 and 1.1 million workers. Illinois and Arizona follow with 560, 000 and 551,000 Hispanic workers. The table also shows that Hispanic

TABLE E1 Employment by Gender, Race, Ethnicity, and Major Industry Group, 1998-2000 (Annual Average of All U.S. Civilian Workers Age 16 and Older)

	Number				Percent			
	Total	White Non-Hispanic	Black Non-Hispanic	Hispanic	Total	White Non-Hispanic	Black Non-Hispanic	Hispanic
Total	133,387	99,053	14,563	13,834	100.0	74.3	10.9	10.4
Men								
Total	71,477	53,333	6,803	8,187	100.0	74.6	9.5	11.5
Agriculture, forestry, and fishing	2,582	1,765	108	654	100.0	68.3	4.2	25.3
Mining	494	416	21	46	100.0	84.3	4.3	9.3
Construction	8,114	6,244	512	1,178	100.0	76.9	6.3	14.5
Durable goods	8,936	6,914	754	858	100.0	77.4	8.4	9.6
Nondurable goods	4,811	3,418	532	676	100.0	71.1	11.1	14.1
Transportation and public utilities	6,786	4,899	953	666	100.0	72.2	14.0	9.8
Wholesale trade	3,623	2,823	253	411	100.0	77.9	7.0	11.3
Retail trade	10,884	7,737	1,032	1,529	100.0	71.1	9.5	14.0
Finance, insurance, and real estate	3,625	2,903	294	267	100.0	80.1	8.1	7.4
Service industries	18,230	13,634	1,900	1,656	100.0	74.8	10.4	9.1
Private household	80	44	12	19	100.0	55.4	14.5	23.4
Public administration	3,313	2,534	432	228	100.0	76.5	13.0	6.9
Women								
Total	61,909	45,721	7,760	5,647	100.0	73.9	12.5	9.1
Agriculture, forestry, and fishing	878	742	19	98	100.0	84.5	2.2	11.2
Mining	75	61	5	6	100.0	81.0	7.2	7.9
Construction	866	742	50	54	100.0	85.7	5.8	6.2
Durable goods	3,403	2,455	375	348	100.0	72.1	11.0	10.2
Nondurable goods	3,098	2,050	409	458	100.0	66.2	13.2	14.8
Transportation and public utilities	2,748	1,919	476	231	100.0	69.8	17.3	8.4
Wholesale trade	1,611	1,232	118	181	100.0	76.5	7.4	11.2
Retail trade	11,419	8,612	1,148	1,109	100.0	75.4	10.1	9.7
Finance, insurance, and real estate	5,090	3,928	600	353	100.0	77.2	11.8	6.9
Service industries	29,228	21,739	3,890	2,375	100.0	74.4	13.3	8.1
Private household	854	446	126	251	100.0	52.2	14.8	29.4
Public administration	2,640	1,796	542	182	100.0	68.0	20.5	6.9

SOURCE: Generated from the microdata of the Current Population Survey, 1998-2000.

TABLE E2 Employment of Hispanics by State and Major Industry Group, 1998-2000 (Annual Average of All Civilian Workers Age 16 and Older)

	California		Florida		New York		Texas	
	Hispanic Employment (000s)	Percent Hispanic by Industry	Hispanic Employ-ment (000s)	Percent Hispanic by Industry	Hispanic Employ-ment (000s)	Percent Hispanic by Industry	Hispanic Employ-ment (000s)	Percent Hispanic by Industry
Total	4,273.9	27.1	1,250.9	17.7	1,070.8	12.7	2,790.3	28.6
Men								
Total	2,566.3	29.5	731.0	19.2	599.2	13.4	1,653.9	30.6
Agriculture, forestry and fishing	307.8	71.1	45.2	31.2	9.6	12.7	99.7	41.1
Mining	2.8	12.1	0.0	0.0	0.0	0.0	28.0	24.1
Construction	300.3	34.9	111.4	21.5	52.3	12.3	316.3	47.8
Durable goods	301.8	30.1	50.1	18.6	29.2	7.8	148.7	27.7
Nondurable goods	207.3	44.2	28.4	20.7	41.8	18.1	139.0	37.7
Transportation and public utilities	194.4	25.8	86.4	22.2	58.8	12.0	148.7	25.4
Wholesale trade	134.4	30.1	49.1	24.1	27.5	14.0	75.2	27.3
Retail trade	469.2	34.9	132.1	18.9	163.1	22.4	289.5	34.7
Finance, insurance and real estate	62.9	14.5	33.3	15.3	46.9	12.7	49.4	20.3
Service industries	519.2	20.3	173.4	17.0	152.5	11.5	302.3	23.0
Private household	8.1	45.6	1.2	22.7	1.5	30.6	4.3	60.3
Public administration	58.0	16.6	20.5	10.8	15.9	6.4	52.7	24.3
Women								
Total	1,707.6	24.1	519.9	15.9	471.6	12.0	1,136.4	26.1
Agriculture, forestry and fishing	54.6	45.3	6.8	15.2	0.5	1.9	9.6	12.7
Mining	1.0	48.8	0.0	0.0	0.0	0.0	3.5	10.8
Construction	12.4	13.8	5.3	9.1	2.6	6.9	15.5	20.7
Durable goods	119.9	27.6	21.0	20.5	19.6	12.2	47.4	27.3
Nondurable goods	156.3	46.5	26.4	27.3	42.0	20.2	69.1	39.5
Transportation and public utilities	60.0	19.4	35.8	21.2	25.0	14.1	46.1	19.2
Wholesale trade	67.0	27.8	24.4	26.0	10.4	10.7	24.2	18.9
Retail trade	311.9	26.2	102.1	15.2	76.5	12.9	267.4	31.9
Finance, insurance and real estate	83.1	14.8	48.3	14.7	38.2	10.6	61.6	18.1
Service industries	679.0	20.4	217.4	14.5	217.5	10.6	507.9	24.8
Private household	112.6	64.0	20.1	40.7	28.7	35.1	48.5	60.5
Public administration	49.7	17.4	12.4	8.1	10.5	6.7	35.6	23.3

SOURCE: Generated from the microdata of the Current Population Survey, 1998-2000

men and women tend to work in different industries in different states. For example, Hispanic men account for a high fraction of employment in agriculture, forestry, and fishing, in California, Texas and Illinois but not in New York, while Hispanic women make up a high fraction of female employment in this industry only in California.

From 1998 to 2000 nearly one-half of all Hispanic men were employed in blue-collar jobs, that is, in the summary occupational groups "precision production, craft, and repair workers" and "operators, fabricators, and laborers" (see Table F). In contrast, nearly 64 percent of Hispanic women were in "technicians, sales, or administrative support" jobs (mostly in administrative support) or in service occupations.

Hispanic men and women were more likely than white workers to be employed in riskier blue-collar and service occupations. Hispanic men accounted for 25.1 percent of male employment in farming, forestry and fishing jobs; 17.4 percent of male employment in service occupations; and 16.4 percent of the occupational group "operators, fabricators, and laborers." In the latter group Hispanic men were especially numerous in the minor occupational group "handlers, equipment cleaners, helpers, and laborers." This group includes occupations requiring less skill, and as seen later, higher job risk.

Black men also were more likely than white men to be employed in blue-collar and service jobs, particularly in service occupations and as operators, fabricators, and laborers. In contrast to Hispanic males, black men are less frequently employed (relative to their percentage in all jobs) in precision, production, craft and repair occupations, and in farming, forestry, and fishing occupations. Even within the summary occupational groups where both black and Hispanic men are over-represented, these two race ethnicity groups tend to be found working in different proportions in the constituent major occupational groups. For example, among the services occupations black men are more likely to be found in protective service occupations, while Hispanic men are more often employed in other service occupations. Similar differences are found in the summary occupational group "operators, fabricators, and laborers."

Similarly, Hispanic women were more frequently found in blue-collar and service jobs. They accounted for 17.5 percent of female employment in the summary occupational group "operators, fabricators, and laborers." Further, they accounted for 15.3 percent of female employment in farming, forestry, and fishing occupations and 13.9 percent in service occupations. Thirty-one percent of female employment in the relatively small private household occupational group was composed of Hispanics.

Compared to white women, black women were also more likely to be found in blue-collar and service occupations. Unlike Hispanic women, they were not often found in farming, forestry, and fishing occupations. Black women and Hispanic women tended to work in different major occupational groups. For example, among operators, fabricators, and laborers, black women make up a high percentage of female workers in transportation and material moving occupations, while Hispanic women make up a higher percentage of machine operators, assemblers, and inspectors. Among service occupations, Hispanic women are over-represented in private household occupations, while black women are over-represented in protective service jobs. The bulk of employment in the service occupation summary group is in "other service occupations," where both black and Hispanic women make up higher fractions than they do of all jobs.

Table G lists the top 10 detailed occupations in terms of employment for Hispanic men and women. No single occupation dominates either list. However, five of the occupations for Hispanic men stand out as generally involving physical labor and low skill: farm workers, janitors and cleaners, groundskeepers and gardeners, and construction and non-construction laborers. Three of the occupations where Hispanic women are most frequently found include cleaning: janitors and cleaners, private household cleaners, and maids. Hispanic women are also found in such traditional lower-skilled "female" jobs as cooks, cashiers, and secretaries.

TABLE F Employment by Gender, Race, Hispanic Ethnicity and Occupational Group, 1998-2000 (Annual Average of All Civilian Workers Age 16 and Older)

	Employment (000s)				Percent			
	Total	White Non-Hispanic	Black Non-Hispanic	Hispanic	Total	White Non-Hispanic	Black Non-Hispanic	Hispanic
Total	133,387	99,053	14,563	13,834	100.0	74.3	10.9	10.4
Men								
Total	71,477	53,333	6,803	8,187	100.0	74.6	9.5	11.5
Managerial and professional specialty	20,285	16,951	1,226	999	100.0	83.6	6.0	4.9
Executive, administrative, and managerial	10,714	9,121	596	575	100.0	85.1	5.6	5.4
Professional specialty	9,571	7,830	630	424	100.0	81.8	6.6	4.4
Technical, sales, and administrative support	14,053	10,844	1,263	1,228	100.0	77.2	9.0	8.7
Technicians and related support	2,063	1,604	179	140	100.0	77.8	8.7	6.8
Sales	8,052	6,559	526	615	100.0	81.5	6.5	7.6
Administrative support, including clerical	3,939	2,681	558	473	100.0	68.1	14.2	12.0
Service occupations	7,187	4,371	1,178	1,251	100.0	60.8	16.4	17.4
Private household	40	21	5	9	100.0	52.9	13.5	23.4
Protective service	1,970	1,418	329	171	100.0	72.0	16.7	8.7
Other service occupations	5,176	2,932	843	1,071	100.0	56.6	16.3	20.7
Precision production, craft, and repair	13,342	10,223	972	1,742	100.0	76.6	7.3	13.1
Mechanics and repairers	4,617	3,629	345	483	100.0	78.6	7.5	10.5
Construction trades	5,700	4,324	385	872	100.0	75.9	6.8	15.3
Other precision production, craft, and repair	3,026	2,271	242	388	100.0	75.0	8.0	12.8
Operators, fabricators, and laborers	13,850	9,077	2,023	2,274	100.0	65.5	14.6	16.4
Machine operators, assemblers, and inspectors	4,713	3,057	629	817	100.0	64.8	13.3	17.3
Transportation and material moving	4,930	3,427	763	612	100.0	69.5	15.5	12.4
Handlers, equipment cleaners, helpers, and laborers	4,207	2,593	631	844	100.0	61.6	15.0	20.1
Farming, forestry and fishing	2,761	1,866	141	693	100.0	67.6	5.1	25.1

Women								
Total	61,909	45,721	7,760	5,647	100.0	73.9	12.5	9.1
Managerial and professional specialty	19,812	15,984	1,892	1,004	100.0	80.7	9.5	5.1
Executive, administrative, and managerial	8,756	7,038	831	491	100.0	80.4	9.5	5.6
Professional specialty	11,056	8,946	1,061	513	100.0	80.9	9.6	4.6
Technical, sales, and administrative support	24,908	18,806	2,997	2,098	100.0	75.5	12.0	8.4
Technicians and related support	2,271	1,721	278	149	100.0	75.8	12.2	6.5
Sales	8,051	6,144	855	684	100.0	76.3	10.6	8.5
Administrative support, including clerical	14,586	10,940	1,864	1,265	100.0	75.0	12.8	8.7
Service occupations	10,823	6,908	1,947	1,499	100.0	63.8	18.0	13.9
Private household	783	405	107	243	100.0	51.7	13.7	31.0
Protective service	449	270	133	33	100.0	60.2	29.6	7.4
Other service occupations	9,592	6,233	1,707	1,223	100.0	65.0	17.8	12.8
Precision production, craft, and repair	1,287	862	160	171	100.0	67.0	12.5	13.3
Mechanics and repairers	226	163	32	18	100.0	71.9	14.3	8.1
Construction trades	139	108	13	14	100.0	78.0	9.1	10.0
Other precision production, craft, and repair	922	591	115	139	100.0	64.1	12.5	15.0
Operators, fabricators, and laborers	4,397	2,622	744	771	100.0	59.6	16.9	17.5
Machine operators, assemblers, and inspectors	2,785	1,554	475	556	100.0	55.8	17.1	20.0
Transportation and material moving	549	386	109	41	100.0	70.4	19.9	7.5
Handlers, equipment cleaners, helpers, and laborers	1,064	682	159	174	100.0	64.1	15.0	16.3
Farming, forestry, and fishing	682	539	19	104	100.0	79.0	2.9	15.3

SOURCE: Generated from the microdata of the Current Population Survey, 1998-2000.

TABLE G Detailed Occupations with the Most Hispanic Employment, by Gender, 1998-2000 (Annual Average of All Civilian Workers Age 16 and Older)

Men	Hispanic Employment (000s)	Percent of Hispanic Employment (000s)	Women	Hispanic Employment (000s)	Percent of Hispanic Employment
All workers	8,187	100.0	All workers	5,647	100.0
Truck drivers	369	4.5	Cashiers	287	5.1
Farm workers	312	3.8	Secretaries	211	3.7
Cooks	310	3.8	Janitors and cleaners	185	3.3
Janitors and cleaners	284	3.5	Private household cleaners	181	3.2
Managers and administrators, not elsewhere classified	269	3.3	Nursing aides, orderlies, and attendants	177	3.1
Groundskeepers and gardeners, except farm	266	3.2	Maids	148	2.6
Construction laborers	232	2.8	Cooks	132	2.3
Supervisors and proprietors, sales occupations	220	2.7	Managers and administrators, not elsewhere classified	128	2.3
Carpenters	209	2.5	Receptionists	112	2.0
Laborers, except construction	181	2.2	Supervisors and proprietors, sales occupations	112	2.0

SOURCE: Generated from the microdata of the Current Population Survey, 1998-2000.

FATAL WORK INJURIES INVOLVING HISPANICS

Overview of National Fatality Data

From 1995 through 2000, 4,167 Hispanics workers died as a result of a fatal injury on the job, accounting for about 11 percent of the total number of fatal work injuries that occurred over this period (see Table H). The number of fatal work injuries involving Hispanic workers has risen each year since 1995, from a low of 619 fatal work injuries in 1995 to a high of 815 fatal work injuries in 2000.

Fatal work injury rates for Hispanic workers have ranged from a low of 5.1 fatal work injuries per 100,000 Hispanic workers in 1997 to a high of 5.6 per 100,000 in 2000. The fatal work injury rate in 2000 for Hispanic workers was 33 percent higher than the rate for non-Hispanic workers and rates for Hispanic workers have been consistently higher than the overall national fatality rate for the period covered by this study.

Event

The most frequent types of fatal event in fatal injuries involving Hispanics were transportation incidents (34 percent) followed by assaults and violent acts (19 percent), contact with objects or equipment (17 percent), and falls (16 percent). The percentages of falls and homicides for Hispanics represented a higher proportion of total fatalities than for all workers. Transportation incidents represented a smaller proportion of total injuries for Hispanics (see Table K).

TABLE H Numbers and Rates Per 100,000 of Fatal Occupational Injuries for Hispanic Workers and Non-Hispanic Workers, United States, 1995-2000

Year	Hispanic Workers		Non-Hispanic Workers	
	Number	Rate	Number	Rate
2000	815	5.6	5068	4.2
1999	730	5.2	5292	4.4
1998	707	5.2	5314	4.5
1997	658	5.1	5561	4.8
1996	638	5.3	5535	4.8
1995	619	5.4	5628	4.9

SOURCE: Bureau of Labor Statistics, Census of Fatal Occupational Injuries.

Industry

The combined count of fatal work injuries involving Hispanic workers in private industry construction (1,153 fatalities) and agriculture (625 fatalities) accounted for 43 percent of the fatal work injuries involving Hispanic workers (see Table IA). Hispanic worker fatalities in transportation and public utilities (493 fatalities) and services (483 fatalities) combined accounted for another 23 percent of the cases. Mining recorded the highest fatal work injury rate of any major industry (37.6 per 100,000), followed by construction (18.3 per 100,000) and agriculture (15.3 per 100,000). Nearly 80 percent of the fatal work injuries in mining occurred in oil and gas extraction industries. Special trade contractors recorded the highest percentage of fatalities in construction (63 percent), and in agriculture the highest percentage of fatal work injuries occurred in agricultural services (42 percent). The number of fatal work injuries to Hispanic workers increased by 32 percent over the period covered by the study, led by fatal work injuries in the construction industry, which nearly doubled over this period (see Table IB).

TABLE IA Number, Percent Distribution, and Rate of Fatal Occupational Injuries Involving Hispanic Workers by Industry Division, 1995-2000

	Number	Percent	Rate	Primary Fatal Injury Event
	4,167	100%	5.4	Highway incidents (19%)
Agriculture, forestry, fishing	625	15	15.3	Highway incidents (22 %)
Mining	109	3	37.6	Highway incidents (27%)
Construction	1,153	28	18.3	Falls to lower level (37%)
Manufacturing	417	10	3.1	Struck by object (20%)
Transportation	493	12	10.1	Highway incidents (40%)
Wholesale trade	160	4	4.9	Highway incidents (23%)
Retail trade	432	10	3	Homicide (68%)
Finance, insurance, and real estate	59	1	1.7	Homicide (31%)
Services	483	12	2	Homicide (28%)
Government	206	5	8.8	Highway incidents (31%)

SOURCE: Bureau of Labor Statistics, Census of Fatal Occupational Injuries.

TABLE IB Number of Fatal Occupational Injuries Involving Hispanic Workers by Detailed Industry, 1995-2000

Industry	1995-2000	1995	1996	1997	1998	1999	2000
Total	4,167	619	638	658	707	730	815
Private industry	3,961	578	600	624	678	698	783
Agriculture, forestry, and fishing	625	103	78	99	114	119	112
Agricultural production - crops	226	41	28	39	43	50	25
Agricultural production - livestock	94	18	13	14	14	18	17
Agricultural services	262	40	34	41	45	37	65
Mining	109	21	18	19	22	14	15
Oil and gas extraction	86	18	18	13	17	10	10
Construction	1,153	142	133	166	211	224	277
General building contractors	172	22	20	22	44	26	38
Heavy construction, except building	235	21	25	38	44	50	57
Special trades contractors	727	99	85	105	122	142	174
Manufacturing	417	55	63	58	78	73	90
Food and kindred products	70	-	16	8	15	11	16
Lumber and wood products	74	7	9	14	13	9	22
Stone, clay, glass, and concrete products	41	-	7	9	5	7	9
Primary metal industries	27	-	-	-	7	7	-
Fabricated metal products	33	-	6	6	5	5	8
Transportation equipment	34	-	-	6	9	8	-
Transportation and public utilities	493	72	82	79	75	81	104
Local and inter-urban passenger transportation	71	16	5	9	7	15	19
Trucking and warehousing	265	38	43	45	47	44	48
Transportation by air	26	-	5	-	5	6	6
Electric, gas, and sanitary services	67	10	12	12	7	7	19
Wholesale trade	160	21	28	41	25	20	25
Durable goods	96	12	18	29	15	10	12
Nondurable goods	64	9	10	12	10	10	13
Retail trade	432	88	87	64	61	65	67
Food stores	124	32	26	14	16	18	18
Automotive dealers and service stations	49	11	-	15	6	10	-
Eating and drinking places	148	21	30	23	23	24	27
Miscellaneous retail	45	10	15	6	6	-	-
Finance, insurance, and real estate	59	11	8	10	11	12	7
Real estate	46	8	7	9	7	9	6
Services	483	62	96	82	76	84	83
Personal services	27	-	7	7	-	8	-
Business services	167	26	29	24	33	28	27
Automotive repair, services, and parking	94	11	13	16	16	20	18
Miscellaneous repair services	34	-	7	5	7	5	7
Amusement and recreation services	44	7	8	7	-	9	9
Government(1)	206	41	38	34	29	32	32
Federal government (including resident armed forces)	70	12	14	10	11	13	10
State government	35	5	5	10	-	8	-
Local government	99	23	19	14	14	11	18

SOURCE: United States Department of Labor, Bureau of Labor Statistics, Census of Fatal Occupational Injuries.

Occupation

More than two out of every five fatally injured Hispanic workers were employed in the operator, fabricator, and laborer occupational summary group at the time of their injuries. Two detailed occupations within that summary group—construction laborers (490 fatalities) and truck drivers (413 fatalities) —accounted for over 20 percent of all fatal work injuries among Hispanic workers and were also the two occupations recording the highest number of fatal injuries (see Table J).

TABLE J Number and percent distribution of fatally injured Hispanic workers by occupation, United States, 1995-2000

	Number	Percent
Total # Hispanic workers	4,167	100
Operators, fabricators, and laborers	1,745	42
Handlers, equipment cleaners, helpers, laborers	882	21
Construction laborers	490	12
Laborers, excluding construction	239	6
Transportation and material moving	652	16
Motor vehicle operators	512	12
Truck drivers	413	10
Precision production	822	20
Construction trades	519	12
Roofers	105	3
Carpenters, apprentices	92	2
Mechanics, repairers	178	4
Vehicle, mobile equipment repair	88	2
Farming, forestry, fishing	626	15
Other agriculture related occupation	547	13
Farm workers	343	8
Groundskeepers, gardeners	143	3
Service occupations	356	9
Service occupations, excluding protective, household	206	5
Cleaning and building service occupations except household	106	3
Protective service occupations.	141	3
Guards	79	2
Technical, sales, administrative support	326	8
Sales occupations	240	6
Managerial, professional specialty	211	5
Executive, administrative, management	139	3
Military occupations	45	1

NOTES:

1. Totals may include categories not shown separately.
2. Percentages may not sum to total due to rounding.

SOURCE: Bureau of Labor Statistics, Census of Fatal Occupational Injuries.

Demographics of Fatally Injured Hispanic Workers

Gender

Male Hispanic workers accounted for 94 percent of the 4,167 fatal work injuries recorded for Hispanics over this period. In terms of event, the highest percentage of fatal work injuries for Hispanic men were highway incidents (18 percent) and falls to lower level (16 percent). For Hispanic women the primary fatal event was workplace homicide (42 percent), followed by highway incidents (25 percent). Table K presents the distribution of fatal work injuries by gender and major event.

The largest percentage of fatally injured Hispanic women were employed in the technical, sales, and administrative support occupations and service occupations at the time of their fatal incidents. The total number of fatal work injuries in these two occupational groups accounted for nearly 6 in 10 of the recorded fatalities. Sales occupations accounted for 45 of the 75 fatal work injuries in technical, sales, and administrative support. Service occupations, except protective and household (e.g., food preparation workers, cleaning and building service workers) recorded 53 of the 70 fatal work injuries in service occupations.

TABLE K Percent Distribution of Fatal Occupational Injuries Involving Hispanic Workers by Gender and Event, United States, 1995-2000

	All Hispanic		
	All Workers	Male	Female
Total number	4,167	3,919	248
	100%	100%	100%
Transportation incidents	34	33	36
Highway incidents	19	18	25
Struck by vehicle, mobile equipment	7	7	6
Non-highway incidents	4	5	--
Assaults and violent acts	19	17	44
Homicides	16	14	42
Self-inflicted injury	3	3	--
Contact with objects or equipment	17	18	5
Struck by object	9	10	3
Caught in or compressed	5	5	--
Caught in or crushed in collapsing materials	3	3	--
Falls	16	16	7
Falls to lower level	15	16	5
Exposure to harmful substances or environments	11	11	6
Electrocution	6	6	--
Fires and explosions	3	3	3
Fires	2	2	--
Explosion	1	1	--

NOTES:
1. Totals may include categories not shown separately.
2. Percentages may not sum to total due to rounding.

SOURCE: Bureau of Labor Statistics, Census of Fatal Occupational Injuries.

The largest percentage of male Hispanic workers who were fatally injured on the job was in the operator, fabricator, and laborer occupations group (44 percent). The combined number of fatalities in two occupations in that occupational group—construction laborers (489 fatalities) and truck drivers (409 fatalities)—accounted for more than one out of every five fatal work injuries involving male Hispanic workers over this period. Another 817 male Hispanic workers were fatally injured while working in precision production, craft, and repair occupations, including 517 in construction trades. Fatal work injuries in the farming, forestry, and fishing occupational group claimed 610 male Hispanic workers, including 333 farm workers.

Employment Status

Hispanic workers categorized as wage and salary workers were involved in nearly 9 in 10 fatal work injuries—a higher percentage than the percentage recorded for all fatally injured workers over this same period (80 percent).

Age

Hispanics are on the whole younger than non-Hispanic whites. According to the Census Bureau the median age for Hispanics in 2000 was 25.9 years, as compared to a median age of 35.3 years for the overall United States population (Census Bureau 2001 [1]). The fatal work injury data results reflect this demographic pattern. Nearly half of the fatally injured Hispanic workers were under 34 years of age, as compared to about 3 in 10 for all fatally injured workers.

Fatality Risk

To assess the fatality risks faced by Hispanics and to evaluate the extent the distribution of employment impacts the risks faced by Hispanics, relative risk measures were calculated by gender, race/ethnicity and by both occupation and industry. Table L reports fatal relative risks for occupational groups. All the risks are relative to the fatality rate for all civilian workers. The table shows that a man is far more likely to suffer a fatal occupational injury than a woman, with men's relative risk being nearly 9 times the value for women. The table also shows that at 1.94, Hispanic men have the highest overall relative risk of fatal occupational injury of any gender or race/ethnicity group in the table. Further, Hispanic men have a relative risk that is 22 percent higher than the relative risk for all men. Hispanic women have a relative risk that is comparable to relative risks faced by white women.

Hispanic men are at especially high risk of fatal work injury in the following occupational groups: handlers, equipment cleaners, helpers, and laborers (4.61 relative risk), transportation and material moving occupations (4.13 relative risk), and farming, forestry, and fishing occupations (3.58 relative risk). Relative risks are especially high for Hispanic women in transportation and material moving occupations and farming, forestry, and fishing occupations.

Table M reports fatal relative risks by industry. Hispanic men have especially high relative risks in mining (7.54), construction (4.76) and agriculture, forestry and fishing (3.75). These industries have high relative risks for all men, but the relative risks in mining and construction are higher for Hispanic men than they are for all men. Hispanic women also have high relative risks of fatality in construction and agriculture, forestry, and fishing.

TABLE L Relative Risk of Fatal Occupational Injury by Gender, Race/Hispanic Ethnicity and Major Occupation, 1998-2000 (All Civilian Workers Age 16 and Older; All Workers = 1.00)

	Men				Women			
	All	White Non-Hispanic	Black Non-Hispanic	Hispanic	All	White Non-Hispanic	Black Non-Hispanic	Hispanic
Total	1.59	1.55	1.68	1.94	0.18	0.18	0.15	0.18
Managerial and professional specialty	0.51	0.50	0.52	0.56	0.13	0.14	0.09	0.14
Executive, administrative, and managerial	0.61	0.59	0.73	0.62	0.13	0.14	0.09	0.10
Professional specialty	0.38	0.39	0.31	0.48	0.13	0.14	0.10	0.19
Technical, sales, and administrative support	0.78	0.76	0.67	0.66	0.14	0.14	0.12	0.15
Technicians and related support	1.64	1.90	0.59	1.22	0.13	0.15	-	-
Sales	0.78	0.65	1.06	0.91	0.23	0.20	0.28	0.24
Administrative support, including clerical	0.30	0.32	0.31	0.14	0.10	0.10	0.06	0.11
Service occupations	1.22	1.38	1.09	0.91	0.18	0.17	0.17	0.18
Private household	-	-	-	-	0.14	-	-	-
Protective service	2.63	2.72	2.23	2.65	0.73	0.97	-	-
Other service occupations	0.60	0.58	0.60	0.62	0.15	0.13	0.17	0.17
Precision production, craft, and repair	1.75	1.71	1.63	2.12	0.21	0.24	-	0.24
Mechanics and repairers	1.45	1.39	1.76	1.64	0.21	0.29	-	-
Construction trades	2.41	2.33	2.23	2.88	0.59	0.52	-	-
Other precision production, craft, and repair	1.06	1.11	0.58	1.09	0.16	0.18	-	-
Operators, fabricators, and laborers	3.23	3.33	3.03	3.17	0.48	0.62	0.37	0.23
Machine operators, assemblers, and inspectors	0.94	0.98	0.75	1.03	0.15	0.18	0.14	0.09
Transportation and material moving	5.11	5.40	4.57	4.13	2.24	2.68	1.05	1.92
Handlers, equipment cleaners, helpers, and laborers	3.52	3.18	3.43	4.61	0.54	0.59	0.67	0.30
Farming, forestry and fishing	6.08	6.80	7.62	3.58	1.00	1.05	-	0.82

NOTE: Relative risks for "other non-Hispanics" not reported.
SOURCE: Calculated from the Bureau of Labor Statistics Census of Fatal Occupational Injuries and the Current Population Survey.

TABLE M Relative Risk of Fatal Occupational Injury by Gender, Race/Hispanic Ethnicity and Major Industry Division, 1998-2000 (All Civilian Workers Age 16 and Older; All Workers = 1.00)

	Men				Women			
	All	White Non-Hispanic	Black Non-Hispanic	Hispanic	All	White Non-Hispanic	Black Non-Hispanic	Hispanic
Total	1.59	1.55	1.68	1.94	0.18	0.18	0.15	0.18
Agriculture, forestry, and fishing	5.64	6.16	7.06	3.75	0.82	0.82	-	0.80
Mining	5.25	5.02	5.36	7.54	-	-	-	-
Construction	3.17	2.85	3.77	4.76	0.49	0.47	-	0.83
Durable goods	1.05	1.02	1.46	1.23	0.10	0.10	0.15	0.14
Nondurable goods	0.81	0.78	1.05	0.88	0.20	0.21	0.22	0.14
Transportation and public utilities	2.91	2.92	2.97	2.87	0.45	0.53	0.24	0.40
Wholesale trade	1.20	1.14	1.77	1.21	0.18	0.20	-	-
Retail trade	0.95	0.84	1.16	0.84	0.22	0.20	0.25	0.22
Finance, insurance and real estate	0.44	0.41	0.51	0.73	0.08	0.09	-	-
Service industries	0.79	0.76	0.93	1.01	0.13	0.12	0.11	0.14
Private household	-	-	-	-	0.12	-	-	-
Public administration	1.82	1.97	1.43	1.21	0.30	0.35	0.20	0.23

NOTE: Relative risks for "other non-Hispanics" not reported.
SOURCE: Calculated from the Bureau of Labor Statistics Census of Fatal Occupational Injuries and the Current Population Survey.

It was noted previously that Hispanics tend to be employed in blue-collar and service occupations. Table L shows that these jobs tend to have higher relative risks for workplace fatality. To what extent do the overall higher relative risks for Hispanic men reflect where they work?

Table N compares standardized and unstandardized relative risks of fatality for gender and race/ethnicity groups. Standardization at the major occupational group level has different effects for Hispanic men and for Hispanic women. Standardization lowers the relative risks for men, but it has the most effect on the relative risk for Hispanic men. After standardization Hispanic men have lower relative risks than white and black non-Hispanics, (though the relative risks for all groups are quite close). This suggests that the higher unstandardized overall relative risk for Hispanic men is largely the result of Hispanic men working in jobs with higher fatality risk.

In contrast to its impact on men, standardization raises the relative risks of women indicating that women tend to work in jobs that are relatively safe in terms of fatality risk. Further, standardization raises relative risk most for white non-Hispanic and Hispanic women, so that both groups of women have the highest relative risk of fatality among all women after accounting for the distribution of hours worked.

TABLE N Standardized and Unstandardized Relative Risks of Fatal Occupational Injury, 1998-2000

	Men	
	Standardized	Unstandardized
White Non-Hispanic	1.25	1.55
Black Non-Hispanic	1.23	1.68
Hispanic	1.20	1.94
	Women	
	Standardized	Unstandardized
White Non-Hispanic	0.34	0.18
Black Non-Hispanic	0.19	0.15
Hispanic	0.34	0.18

NOTE: Standardization by major occupational group. Relative risks for "other non-Hispanics" not reported.
SOURCE: Bureau of Labor Statistics, Census of Fatal Occupational Injuries.

Foreign-Born Hispanic Workers

Foreign-born workers appear to bear a disproportionate share of the fatal work injury burden among Hispanic workers. Of the 4,167 fatal work injuries involving Hispanic workers from 1995 through 2000, the CFOI found that 2,440 of those fatalities (or 59 percent) involved workers who were born outside of the United States. Moreover, the rate for foreign-born Hispanic (1996 through 2000) was 6.1 per100,000 as compared to a rate of 4.5 per100,000 for native-born Hispanic workers. The rate for all workers over this same period was 4.6 per 100,000.

Workers born in Mexico accounted for the majority of these fatally injured, foreign-born workers, with 1,682 cases recorded, or 69 percent of the total. Fatally injured workers from Cuba (146 fatal work injuries), El Salvador (131), Guatemala (90), and Dominican Republic (87) recorded the next highest totals.

Table O presents the five most frequent countries of birth for fatally injured, foreign-born workers, along with the primary state where these events occurred and the primary fatal event associated with each group.

A list of the specific occupations with the highest numbers of fatal work injuries involving Hispanic workers appears in Table Q, along with the proportion of those fatalities that involved foreign-born workers.

Foreign-born workers who were fatally injured on the job had a higher percentage of fatalities from falls and contact with objects and equipment, and a lower percentage of fatalities from transportation incidents than native-born Hispanics. For foreign-born workers the combined number of fatal work injuries in agriculture, forestry, and fishing and in construction accounted for half of all fatal work injuries for that population, as compared to about a third of the fatal work injuries to native-born Hispanic workers. Table P presents a comparative distribution of fatal work injuries by fatal event and industry for all Hispanic workers, foreign-born Hispanic workers, and native-born Hispanic workers.

A list of the specific occupations with the highest numbers of fatal work injuries involving Hispanic workers appears in Table Q, along with the proportion of those fatalities that involved foreign-born Hispanic workers.

TABLE O Fatal Work Injuries Involving Foreign-born Hispanic Workers by Country of Birth, Primary State of Injury, and Primary event, U.S., 1995-2000

	Number	Percent	Primary States	Primary fatal event
Total Foreign-Born	2,440	100	CA, TX, FL	Homicide (16%)
Mexico	1,682	69	CA, TX	Falls to lower level (18%)
Cuba	146	6	FL	Homicide (32%)
El Salvador	131	5	CA, TX	Homicide (21%)
Guatemala	90	4	CA, FL	Highway incidents, Homicides, Falls to lower level (17% each)
Dominican Republic	87	4	NY, PA	Homicide (72%)

NOTES
1. Totals may include categories not shown separately.
2. Percentages may not sum to total due to rounding.
SOURCE: Bureau of Labor Statistics, Census of Fatal Occupational Injuries.

TABLE P Percent Distribution of Fatal Work Injuries for All Hispanic Workers, Foreign-born Hispanic Workers, and Native-born Hispanic Workers by Event, U.S., 1995-2000

	All Hispanic	Foreign-born Hispanic	Native-born Hispanic
Number	4,167	2,440	1,727
Percent	100	100	100
Event			
Transportation incidents	34	30	39
Assaults and violent acts	19	19	19
Contact with objects or equip	17	19	15
Falls	16	18	12
Exposure to harmful substances	11	11	10
Fires, explosions	3	3	4
Industry			
Agriculture, forestry, fishing	15	19	9
Mining	3	2	4
Construction	28	31	23
Manufacturing	10	11	9
Transportation, public utilities	12	11	13
Wholesale trade	4	4	4
Retail trade	10	10	11
Finance, insurance, real estate	1	1	2
Services	12	10	14
Government	5	1	10

NOTES: Percentages may not sum to total due to rounding.
SOURCE: Bureau of Labor Statistics, Census of Fatal Occupational Injuries.

TABLE Q Occupations with Highest Number of Fatal Occupational Injuries to Hispanics with Percentage Involving Foreign-born, 1995-2000

Occupation	Number	Percentage Involving Foreign-born
Construction laborers	490	72
Truck drivers	413	49
Farm workers	343	80
Laborers, except construction	239	71
Groundskeepers, gardeners, except farm	143	69
Supervisors, proprietors, sales occupations.	116	61
Roofers	105	70
Carpenters, apprentices	92	73
Janitors, cleaners	88	50
Guards, police, except public service	76	54

SOURCE: Bureau of Labor Statistics, Census of Fatal Occupational Injuries.

State Comparisons

The five states with the largest Hispanic populations (CA, TX, FL, NY, and IL) also had the highest number of fatal injuries to Hispanic workers. Two out of every three fatal work injuries from 1995 through 2000 involving Hispanic workers occurred in these five states. As mentioned previously, about 50 percent of the Hispanics in the United States live in either California or Texas and those two states accounted for nearly the same percentage of the fatal work injuries (49 percent).

As noted earlier, about 60 percent of Hispanic worker deaths over this period involved workers who were born in another country. However, the percentage of fatally injured foreign-born workers varied considerably from state to state (see Table R). California, which recorded the highest number of fatally injured Hispanic workers and the highest number of fatally injured foreign-born Hispanic workers, reported that about three out of five Hispanic worker fatalities over this period involved foreign-born workers. Texas, which recorded the second highest totals in both categories, found that less than half of the Hispanic workers who were killed on the job were foreign-born.

TABLE R Fatal Work Injuries to Native and Foreign-born Hispanic Workers by State and Country of Birth, U.S., 1995-2000

State	Total	Hispanic (% of Foreign-born)	Hispanic Birth for Foreign-born (% of Foreign-born)
California	1,112	675 (61%)	Mexico (85%)
Texas	922	444(48%)	Mexico (92%)
Florida	420	334 (80%)	Cuba (36%)
New York	274	183 (67%)	Dominican Republic (35%)
Illinois	114	85 (76%)	Mexico (91%)

NOTES:
1. Totals may include categories not shown separately.
2. Percentages may not sum to total due to rounding.
SOURCE: Bureau of Labor Statistics, Census of Fatal Occupational Injuries.

The birth country of foreign-born Hispanic workers killed on the job also varied by state. In California, Texas, and Illinois the primary birth country was Mexico. In Florida the primary birth country was Cuba, and in New York it was the Dominican Republic.

The five states with the largest percentage gains in their Hispanic populations over the past decade (see Table C) also recorded higher numbers of fatal work injuries. All five of these states reported highs in the number of Hispanic worker fatalities in 2000: North Carolina (22 fatal work injuries), Arkansas (9), Georgia (26), Tennessee (12), and Nevada (10).

Non-Fatal Injuries Involving Hispanic Workers

After imputing for gender and race/ethnicity as described in the methods section, we estimate that annually there were 182,000 injuries to Hispanic men and 66,000 injuries to Hispanic women over the period 1998 to 2000. These estimates do not include mining industries and railroads, for which it was not possible to impute race/ethnicity. (Table S) Unless otherwise specified, all numbers reported in the text are the results after imputation.

TABLE S Non-fatal Occupational Injuries with Days Away from Work to Hispanics, by Gender and Industry group, 1998-2000 (Annual Average of Private Industry Workers Age 16 and Older)

	Men With imputation (000s)	Women With imputation (000s)
Total	182.0	66.4
Agriculture, forestry, and fishing	15.3	2.3
Construction	31.2	-
Durable goods	26.7	6.3
Nondurable goods	17.6	7.4
Transportation and public utilities	20.1	3.4
Wholesale trade	16.2	3.0
Retail trade	24.3	14.5
Finance, insurance, and real estate	3.5	2.2
Service industries	25.4	26.7

NOTES:
1.Totals include industries and values not reported separately. Dash signifies fewer than 500 cases.
2. Excludes SICs 10, 12, 14, and 40 and farms with fewer than 11 employees.
SOURCE: Bureau of Labor Statistics, Survey of Occupational Injuries and Illnesses.

Injuries to Hispanic men were distributed among several major industry divisions. Construction was the division with the most annual injuries (31,200). However, nondurable goods, service industries, retail trade and transportation and public utilities each showed 20,000 or more injuries when race/ethnicity was imputed. In contrast, injuries to Hispanic women were concentrated in the service industries (26,700) and to a lesser extent in retail trade (14,500).

"Operators, fabricators, and laborers" was the summary occupational group with the most injuries to Hispanic men, with 48 percent of that category's injuries in the low-skilled major occupational group "handlers, equipment cleaners, helpers, and laborers" (see Table T). The other blue-collar summary occupational group—precision production, craft and repair occupations—was second in terms of the number of injuries to Hispanic men. Service occupations ranked first in the number of injuries to Hispanic women at 24, 200 with imputation. "Operators, fabricators,

TABLE T Non-fatal Occupational Injuries with Days Away from Work to Hispanics, by Gender and Occupation Group, 1998-2000 (Annual Average of Private Industry Workers Age 16 and Older)

	Men With imputation (000s)	Women With imputation (000s)
Total	182.0	66.4
Managerial and professional specialty	2.5	3.3
Executive, administrative, and managerial	1.5	1.4
Professional specialty	1.0	1.9
Technical, sales, and administrative support	11.6	14.0
Technicians and related support	1.7	1.6
Sales	3.3	5.4
Administrative support, including clerical	6.6	7.0
Service occupations	21.3	24.2
Precision production, craft, and repair	37.3	2.7
Mechanics and repairers	11.5	-
Construction trades	17.9	-
Other precision production, craft, and repair	7.9	2.2
Operators, fabricators, and laborers	91.3	18.7
Machine operators, assemblers, and inspectors	28.1	11.5
Transportation and material moving	19.3	1.1
Handlers, equipment cleaners, helpers, and laborers	43.9	6.1
Farming, forestry, and fishing	16.2	2.5

NOTES:
1. Total includes cases with missing occupation and values not reported separately. Dash signifies fewer than 500 cases.
2. Excludes SICs 10, 12, 14, and 40, and farms with fewer than 11 employees
SOURCE: Bureau of Labor Statistics Survey of Occupational Injuries and Illnesses.

and laborers" and "technical, sales and administrative support" ranked second and third for Hispanic women.

Seven occupations that rank in the top 10 in number of non-fatal injuries and illnesses to Hispanic men also appear in the list of the 10 occupations with the most fatalities. Topping the non-fatal list are laborers—both construction and nonconstruction laborers—and truck drivers (see Table U). Farm workers and groundskeepers and gardeners also appear in the non-fatal list. Topping the list of occupations with the most non-fatal injuries and illnesses to Hispanic women are nursing aides, orderlies, and attendants. Cleaning jobs—maids and janitors and cleaners—rank second and third. Two jobs in the top 10 for Hispanic women involve food preparation.

TABLE U Occupations with the Most Non-fatal Injuries and Illnesses with Days Away from Work to Hispanics, 1998-2000 (Annual Average of Private Industry Workers Age 16 and Older)

Men with Imputation (000s)	
Laborers, except construction	13.9
Truck drivers	13.4
Construction laborers	11.3
Farm workers	7.9
Freight stock and material handlers, nec	7.4
Janitors and cleaners	6.5
Groundskeepers and gardeners	5.8
Miscellaneous machine operators, nec	5.7
Carpenters and apprentices	4.7
Cooks	3.5
Women with Imputation (000s)	
Nursing aides, orderlies, and attendants	6.1
Maids	4.8
Janitors and cleaners	3.1
Laborers, except construction	2.2
Assemblers	2.0
Miscellaneous food preparation occupations	2.0
Miscellaneous machine operators, nec	1.9
Cashiers	1.7
Cooks	1.6
Sales workers, other commodities	1.5

NOTE: Excludes SICs 10, 12, 14, and 40, and farms with fewer than 11 employees.
SOURCE: Calculated from the Bureau of Labor Statistics, Survey of Occupational Injuries and Illnesses and the Current Population Survey.

Relative Risks for non-Fatal Injuries and Illnesses

As with fatalities we calculated relative risks for non-fatal injuries by occupational group and industry division. As discussed in the methods section the scope of the relative risks are private industry workers, excluding agricultural production (SICs 01 and 02), mining (SICs 10, 12, and 14), and railroads (SIC 40). Mining and railroads are excluded because no race/ethnicity information is provided. Agricultural production is excluded because the Survey of Occupational Injuries and Illnesses excluded farms with fewer than 11 workers, and the Current Population Survey does not identify farm size. Thus, the non-fatal relative risks are all relative to the rate of injuries to all private industry workers excluding those industries listed above.

As with fatalities Hispanic men have a higher risk of non-fatal workplace injury or illness than any gender/race/ethnicity group (see Table V). Hispanic men's relative risk is 50 percent higher than for all workers and it is 33 percent higher than the relative risk for all men. The highest relative risk for Hispanic men among occupational groups is handlers, equipment cleaners, helpers, and laborers. However, the relative risk for Hispanic men in that occupational group is lower than the relative risk for all men in that group. Other jobs with high relative risk, both for Hispanic men and for all men, include agricultural service, forestry and fishing jobs, and the other occupational groups in the group "operators, fabricators, and laborers."

TABLE V Relative Risk of Non-fatal Occupational Injury and Illness by Gender, Race/Hispanic Ethnicity and Occupation Group, 1998-2000 (Private Industry Workers Age 16 and Older)

	Men				Women			
	All	White Non-Hispanic	Black Non-Hispanic	Hispanic	All	White Non-Hispanic	Black Non-Hispanic	Hispanic
Total	1.13	1.07	1.40	1.51	0.81	0.76	1.06	1.00
Managerial and professional specialty	0.12	0.11	0.23	0.20	0.31	0.29	0.47	0.33
Executive, administrative, and managerial	0.11	0.11	0.19	0.19	0.18	0.17	0.30	0.23
Professional specialty	0.13	0.13	0.27	0.22	0.47	0.46	0.65	0.48
Technical, sales and administrative support	0.49	0.46	0.69	0.66	0.51	0.49	0.58	0.56
Technicians and related support	0.73	0.75	0.79	0.76	0.88	0.86	1.16	0.83
Sales	0.30	0.29	0.37	0.36	0.59	0.59	0.60	0.61
Administrative support, including clerical	0.86	0.80	1.06	1.08	0.39	0.36	0.47	0.49
Service occupations	1.33	1.37	1.49	1.24	1.84	1.84	1.98	1.76
Precision production, craft and repair	1.43	1.48	1.10	1.43	1.08	1.16	0.94	1.08
Mechanics and repairers	1.46	1.48	1.23	1.59	1.16	1.13	1.21	-
Construction trades	1.70	1.81	1.25	1.44	1.78	1.97	-	-
Other precision production, craft and repair	1.00	0.98	0.76	1.24	0.99	1.07	0.80	1.04
Operators, fabricators, and laborers	2.68	2.81	2.45	2.51	2.16	2.52	1.96	1.61
Machine operators, assemblers, and inspectors	2.22	2.41	1.84	2.04	1.72	2.09	1.45	1.32
Transportation and material moving	2.33	2.46	2.06	2.05	3.06	3.13	3.12	2.62
Handlers, equipment cleaners, helpers, and laborers	3.68	3.86	3.65	3.31	3.14	3.40	3.14	2.48
Agricultural service, forestry and fishing occupations	2.13	2.11	2.03	2.20	1.62	1.60	-	1.78

NOTES:
1. Excludes SICs 01, 02, 10, 12, 14, and 40.
2. Relative risk for all private industry workers except exclusions = 1.0.
3. Agricultural service, forestry, and fishing occupations exclude workers in agricultural production. Dashes signify a relative risk based on fewer than 500 cases. Relative risks for "other non-Hispanics" not reported.
SOURCE: Calculated from Bureau of Labor Statistics, Survey of Occupational Injuries and Illnesses and the Current Population Survey.

TABLE W Relative Risk of Non-fatal Occupational Injury and Illness by Gender, Race/Hispanic Ethnicity and Industry Division, 1998-2000 (Private Industry Workers Age 16 and Older).

	Men				Women			
	All	White Non-Hispanic	Black Non-Hispanic	Hispanic	All	White Non-Hispanic	Black Non-Hispanic	Hispanic
Total	1.12	1.06	1.39	1.50	0.82	0.77	1.07	1.01
Agriculture service, forestry, and fishing	1.69	1.47	1.77	2.11	0.92	0.79	0.95	2.13
Construction	1.70	1.68	1.76	1.80	0.44	0.42	0.66	0.45
Durable goods	1.30	1.26	1.46	1.76	1.01	1.03	1.12	1.11
Nondurable goods	1.11	1.03	1.26	1.48	0.92	0.91	1.01	1.03
Transportation and public utilities	1.67	1.56	2.11	2.16	1.16	1.09	1.60	1.13
Wholesale trade	1.77	1.63	2.84	2.32	0.80	0.74	1.17	1.09
Retail trade	0.94	0.94	1.14	0.98	0.87	0.86	0.94	0.98
Finance, insurance, and real estate	0.32	0.25	0.53	0.84	0.27	0.24	0.34	0.42
Service industries	0.65	0.55	1.01	1.15	0.82	0.73	1.17	1.07

NOTES:
1. Excludes SICs 01, 02, 10, 12, 14, and 40.
2. Relative risk for all private industry workers except exclusions – 1.0.
3. Agricultural service, forestry and fishing excludes workers in agricultural production. Relative risks for "other non-Hispanics" not reported.
SOURCE: Calculated from the Bureau of Labor Statistics, Survey of Occupational Injuries and Illnesses and the Current Population Survey.

Hispanic women have a higher relative risk of non-fatal injury than white women do, but their relative risk is slightly lower than for black women. For both Hispanic women and all women, the highest relative risks are for transportation and material-moving occupations and for handlers, equipment cleaners, helpers, and laborers.

Hispanic men have higher non-fatal relative risks than all men do in each major industry division (see Table W). Since this is not the case by occupational group (compare with Table V), this suggests that Hispanic men tend to work in riskier jobs in each industry division. Similarly, Hispanic women have higher relative risks than all women do in most industry divisions. However, the differences are not as marked for women as for men.

To assess the extent that differences in employment distribution are responsible for the higher non-fatal relative risks of Hispanic men and women, we calculated relative risks that standardize for employment at the major occupational group level. Table X shows that the occupational distribution of employment fully explains aggregate differences in non-fatal risk. After controlling for occupation, Hispanic men have essentially the same relative risk (1.00) as non-Hispanic black and white men. Further, Hispanic women have a lower relative risk than non-Hispanic black and white women do.

TABLE X Standardized and Unstandardized Relative Risks of Non-fatal Occupational Injury and Illness, 1998-2000 (Private Industry Workers Age 16 and Older)

	Standardized	Unstandardized
	Men	
White non-Hispanic	1.02	1.07
Black non-Hispanic	1.00	1.40
Hispanic	1.00	1.51
	Women	
White non-Hispanic	1.08	0.76
Black non-Hispanic	1.10	1.06
Hispanic	0.95	1.00

NOTE: Relative risks for "other non-Hispanics" not reported
SOURCE: Calculated from the Bureau of Labor Statistics, Survey of Occupational Injuries and Illnesses and the Current Population Survey.

Non-fatal Occupational Injury and Illness Characteristics

Besides providing information on the number of cases with days away from work, the Survey of Occupational Injuries and Illnesses provides information on each case, including its nature (e.g., fracture, sprain, etc.), event (e.g. fall, contact with equipment) and the number of days out of work. This section presents information on these characteristics by the gender and race/Hispanic-ethnicity of the injured workers. Unlike the previous section, missing gender and race/Hispanic ethnicity was not imputed. Instead, data are reported separately for injuries with this missing information.

As they are for all workers, sprains and strains are the most common injury to both Hispanic men and women (see Table Y). However, for a Hispanic a particular case is less likely to be a sprain or strain than it is for all workers. Between 1998 and 2000, sprains and strains accounted for 42 percent of cases for all men, but only 35.3 percent of cases for Hispanic men. Similarly, sprains and strains accounted for 47 percent of cases for all women, but only 41.5 percent for Hispanic women. Hispanic men were more likely to sustain cuts and "all other" cases than all men and slightly more likely to sustain fractures, amputations, and bruises and contusions. Hispanic women were more likely to sustain "all other" cases and cuts and lacerations, but slightly less likely to sustain fractures, bruises and contusions, carpal tunnel syndrome, and tendonitis.

The type of event that resulted in a workplace injury or illness in 1998 to 2000 was quite different for a Hispanic worker than for other workers (see Table Z). For all men bodily reaction and exertion (such as lifting, running, slipping, pushing, and repetitive motion) was the leading event, accounting for 41.3 percent of cases. In contrast, this event category accounted for 34.8 percent of cases for Hispanic men. The leading event for Hispanic men was contact with objects and equipment (such as being struck or rubbed by an object or being caught in an object), accounting for 37.6 percent of cases. For all men, contact with objects and equipment accounted for 31.3 percent.

TABLE Y Number and Percent of Workplace Injuries and Illnesses with Days Away from Work by Type, Gender, and Race/Ethnicity, 1998-2000 (Annual Average of Private Industry Workers Age 16 and Older)

	Number of Cases (000s)					Percent of Cases				
	All	White Non-Hispanic	Black Non-Hispanic	Hispanic	Not Reported	All	White Non-Hispanic	Black Non-Hispanic	Hispanic	Not Reported
	Men									
Total	1,116.5	584.0	87.2	136.1	289.2	100.0	100.0	100.0	100.0	100.0
Fractures	83.1	47.4	5.9	10.9	17.3	7.4	8.1	6.8	8.0	6.0
Sprains and strains	469.2	246.6	37.6	48.0	129.6	42.0	42.2	43.1	35.3	44.8
Amputations	8.4	4.6	0.6	1.6	1.3	0.8	0.8	0.7	1.2	0.4
Cuts and lacerations	104.3	52.6	7.5	17.2	24.2	9.3	9.0	8.6	12.6	8.4
Bruises and contusions	94.6	45.9	9.0	12.1	25.9	8.5	7.9	10.3	8.9	9.0
Chemical burns	8.4	4.6	0.9	1.0	1.6	0.8	0.8	1.0	0.7	0.6
Heat burns	17.6	9.2	1.7	2.3	3.8	1.6	1.6	1.9	1.7	1.3
Carpal tunnel syndrome	8.5	5.4	-	-	2.2	0.8	0.9	-	-	0.8
Tendonitis	6.3	3.6	-	0.5	1.6	0.6	0.6	-	0.4	0.6
All other	315.6	163.7	22.8	41.6	81.2	28.3	28.0	26.1	30.6	28.1
	Women									
Total	561.3	270.8	63.0	46.3	167.9	100.0	100.0	100.0	100.0	100.0
Fractures	30.3	17.7	2.4	2.3	6.9	5.4	6.5	3.8	5.0	4.1
Sprains and strains	263.7	129.5	29.9	19.2	79.9	47.0	47.8	47.5	41.5	47.6
Amputations	1.3	0.7	-	-	-	0.2	0.3	-	-	-
Cuts and lacerations	24.5	11.6	2.5	2.5	7.1	4.4	4.3	4.0	5.4	4.2
Bruises and contusions	56.3	24.4	7.4	4.5	18.6	10.0	9.0	11.7	9.7	11.1
Chemical burns	2.3	1.1	-	-	0.5	0.4	0.4	-	-	0.3
Heat burns	8.7	3.9	1.1	0.8	2.3	1.5	1.4	1.7	1.7	1.4
Carpal tunnel syndrome	18.6	10.3	1.8	1.4	4.6	3.3	3.8	2.9	3.0	2.7
Tendonitis	9.5	5.1	0.8	0.6	2.6	1.7	1.9	1.3	1.3	1.5
All other	145.5	66.1	16.3	14.3	44.7	25.9	24.4	25.9	30.9	26.6

NOTES:
1. Excludes SICs 10, 12, 14, and 40.
2. Cases with missing gender and cases for "other non-Hispanics" not reported. Dashes signify fewer than 500 cases.
SOURCE: Bureau of Labor Statistics, Survey of Occupational Injuries and Illnesses.

TABLE Z Number and Percent of Workplace Injuries and Illnesses with Days Away from Work by Event, Gender, and Race/Ethnicity, 1998-2000 (Annual Average of Private Industry Workers Age 16 and Older)

	Number of cases (000s)					Percent of cases				
	All	White Non-Hispanic	Black Non-Hispanic	Hispanic	Not Reported	All	White Non-Hispanic	Black Non-Hispanic	Hispanic	Not Reported
Men										
Total	1116.5	584.0	87.2	136.1	289.2	100.0	100.0	100.0	100.0	100.0
Contact with objects and equipment	349.6	180.0	28.8	51.2	82.3	31.3	30.8	33.0	37.6	28.5
Falls	174.9	94.2	12.3	21.7	43.9	15.7	16.1	14.1	15.9	15.2
Bodily reaction and exertion	460.9	243.9	32.4	47.4	129.8	41.3	41.8	37.2	34.8	44.9
Exposure to harmful substances	48.4	25.6	4.4	5.8	11.4	4.3	4.4	5.0	4.3	3.9
Transportation accidents	53.5	26.0	7.0	5.9	13.5	4.8	4.5	8.0	4.3	4.7
Fires and explosions	3.4	2.0	-	-	0.6	0.3	0.3	-	-	0.2
Assaults and violent acts	9.7	4.5	1.0	1.1	2.6	0.9	0.8	1.1	0.8	0.9
Other events and exposures	15.9	7.5	0.8	2.3	4.8	1.4	1.3	0.9	1.7	1.7
Women										
Total	561.3	270.8	63.0	46.3	167.9	100.0	100.0	100.0	100.0	4.9
Contact with objects and equipment	104.1	48.1	12.6	10.4	30.3	18.5	17.8	20.0	100.0	2.4
Falls	119.1	58.3	13.0	11.2	33.8	21.2	21.5	20.6	22.5	3.2
Bodily reaction and exertion	273.4	135.0	28.7	19.9	83.6	48.7	49.9	45.6	18.0	-
Exposure to harmful substances	26.1	11.6	2.8	2.4	8.3	4.6	4.3	4.4	24.2	-
Transportation accidents	17.4	7.9	2.8	1.1	5.3	3.1	2.9	4.4	20.1	1.1
Fires and explosions	-	-	-	-	-	-	-	-	43.0	2.1
Assaults and violent acts	13.1	6.3	2.1	0.5	3.6	2.3	2.3	3.3	49.8	1.1
Other events and exposures	7.4	3.3	0.6	0.5	2.6	1.3	1.2	1.0	5.2	1.5

NOTES:
1. Excludes SICs 10, 12, 14, and 40.
2. Cases with missing gender and cases for "other non-Hispanics" not reported. Dashes signify fewer than 500 cases annually.
SOURCE: Bureau of Labor Statistics, Survey of Occupational Injuries and Illnesses.

For Hispanic women, as for all women, the leading event was bodily reaction and exertion. However, this category was less frequent for Hispanic women (43 percent) than for all women (48.7 percent). Hispanic women were more likely to be affected by contact with objects and equipment or falls, and slightly more likely to be harmed by exposure to harmful substances. Hispanic women were less likely to suffer from a transportation accident or an assault or other violent act.

Duration of Occupational Injuries and Illnesses

When a Hispanic worker suffers an injury with days away from work, that injury is likely to last longer than for other workers. Between 1998 and 2000 the median days away from work for Hispanic men was 7 days in contrast to 6 days for all men, while the median days away from work for Hispanic women was 6 days, in comparison to 5 days for all women (see Table AA). This disparity is not accounted for by differences in the types of injuries and illnesses. Hispanic men and women tend to be out of work longer for many specific natures of injury. For example, the median days away from work for fractures was 29 days for Hispanic men and 27 days for

Hispanic women, in comparison to 20 days for all men and 18 days for all women. Sprains and strains lasted 7 days for Hispanic men compared to 6 days for all men, and 6 days for Hispanic women compared to 5 days for all women.

There are two possible explanations for the longer durations of injuries and illnesses for Hispanics. One is that Hispanic workers are injured more severely, possibly because they are employed more frequently as blue-collar and service workers, where injuries are more severe. Another is that Hispanics are reluctant to report less severe cases.

TABLE AA Median Number of Days Away from Work, by Type of Injury, Gender, and Race/Ethnicity, 1998-2000 (Private Industry Workers Age 16 and Older).

	All Workers	White Non-Hispanic	Blanc Non-Hispanic	Hispanic
Men				
Total	6	6	5	7
Fractures	20	20	21	29
Sprains and strains	6	6	6	7
Amputations	18	16	25	24
Cuts and lacerations	3	3	3	5
Bruises and contusions	3	3	4	4
Chemical burns	3	3	3	3
Heat burns	5	5	7	6
Carpal tunnel syndrome	27	26	-	-
Tendonitis	9	9	-	9
All other	6	6	5	7
Women				
Total	5	5	5	6
Fractures	18	16	25	27
Sprains and strains	5	5	5	6
Amputations	15	15	-	-
Cuts and lacerations	3	3	3	4
Bruises and contusions	3	3	4	3
Chemical burns	2	2	-	-
Heat burns	3	2	5	4
Carpal tunnel syndrome	26	24	23	35
Tendonitis	10	10	11	18
All other	5	5	5	6

NOTES:
1. Excludes SICs 10, 12, 14, and 40.
2. Based on non-imputed race/ethnicity. Cases with missing gender or race/ethnicity and for "other non-Hispanics" not reported. Dashes signify the median is based on less than 500 annual cases.
SOURCE: Bureau of Labor Statistics, Survey of Occupational Injuries and Illnesses.

Appendix D-1

Background On Methods

The Bureau of Labor Statistics (BLS) conducts two data programs that obtain information on workplace safety and health. The annual Survey of Occupational Injuries and Illnesses (SOII) is a mandatory survey that collects data on non-fatal workplace injuries and illnesses from a stratified random sample of approximately 176,000 private industry establishments. Excluded from the data collection are government agencies, farms with fewer than 11 employees, and the self-employed. These exclusions might be especially problematic when studying Hispanics, since Hispanics have a higher prevalence of employment on farms and in private households. Data for mines and railroads are provided to the BLS from other federal government agencies.

Among the data elements collected for each sampled injury or illness that required one or more days away from work are the gender, age, race/ethnicity, and occupation of the worker, as well as the nature of injury (e.g., sprain, fracture), part of body, source of injury or illness (e.g., tool, surface), and the event (e.g., fall, assault). While race/ethnicity is collected, it is not a required field. As a result, 28 percent of cases have unreported race/ethnicity for the years 1998-2000. Furthermore, race/ethnicity is not available for SICs 10, 12, 14, and 40 and in the SOII because SOII data for these industries come from outside the BLS, and these other sources do not collect race/ethnicity. Weights are attached to each observation so that estimates of the population of all disabling workplace injuries and illnesses can be generated. The data reported through the SOII are based on records that employers maintain under the Occupational Safety and Health Act of 1970 and in some cases from Workers' Compensation records. Because this program is a survey and not a census, it is subject to sampling error.

The BLS's other workplace safety and health data program, the Census of Fatal Occupational Injuries (CFOI), is a federal/state cooperative program administered by the BLS which collects detailed information on all work-related fatalities from injury occurring during a given year (including private wage and salary workers, public sector employees—both civilian and military—and the self-employed).

More than 30 data elements are collected through the CFOI program. Included in the results are such demographic data about as employee work status (wage or salary worker, or self-employed); gender; age; race or ethnic origin; occupation information (classified according to the Bureau of Census's 1990 Occupational Classification System); and employment industry (classified according to the 1987 Standard Industrial Classification System). Other data elements are coded according to the Bureau of Labor Statistics' Occupational Injury and Illness Classification Manual and consist of such circumstances as the event or exposure causing the fatality; sources of the injury; activity of the worker during the time of the incident; and the location where the fatal injury took place. Approximately 7 percent of fatalities occurred to workers whose race was unknown. Fatally injured workers who were born in Puerto Rico but were working in the United States at the time of the fatal incident were considered "native born" for the purposes of this study. Race/ethnicity is not self-declared, which may result in coding errors.

Diverse source documents are used to compile fatality counts that are as complete as possible. Each fatality is typically verified using at least two source documents consisting of death certificates, medical examiner or coroner reports, state Workers' Compensation fatality reports, and other sources that may be available.

Fatality rates reported in Table H are calculated based on annual average employment from the Current Population Survey. Because employment data are not collected by the CFOI, fatality rates were calculated using estimates of employed civilian workers (age 16 and older) from the Current Population Survey (CPS) (described below). Resident military figures, obtained from the Department of Defense (DOD), were added to the CPS employment totals. There are some limitations to these fatality rates: (1) rates are based on employment regardless of hours worked; (2) CPS classifies occupation based on the primary job worked which may not be the job the decedent was performing when fatally injured; and (3) because CPS is a survey rather than a census, sample error may be present in the CPS data. Fatality rates should not be confused with relative fatality risk calculations described below.

Employment and hours worked for worker groups were calculated from the microdata of the CPS. The CPS is a monthly random sample of 50,000 households that represents the entire noninstitutionalized civilian population of the United States. In addition to obtaining demographic information about each worker in surveyed households, the survey asks questions about the worker's industry, occupation, and hours worked per week. We used 36 months of the CPS microdata and a total of 2.16 million records for employed persons to estimate average annual employment and total hours worked for 1998 to 2000. Because the CPS only asks about hours worked in the survey week, these weekly estimates were multiplied by 4.33 to obtain a monthly estimate. Hours-worked estimates for both the main job and a second job, if there was one, were used. Three-year-total hours worked for any group of workers is simply the weighted sum of the monthly hours estimates for all microdata observation in the group, applying sample weights.

Using data from all three programs, we calculated measures of relative risk to assess the risk of workplace injury, illness, and death. Separate relative risk measures were calculated for fatalities and for non-fatal cases. The relative risk for a group of workers is calculated as the injury or fatality rate for that group divided by the injury or fatality rate for all workers. The injury or fatality rate used in these calculations is simply the number of injuries or fatalities sustained by a group of workers during a reference time period divided by the hours worked by that group of workers in the same reference period. The relative risk measures how much the injury risk of a reference group (e.g., Hispanic men) differs from the injury risk of all workers. For example, a relative risk of 1.9 for deaths of Hispanic men indicates that the fatality rate for Hispanic men is 1.9 times the fatality rate for all workers.

We calculated relative fatality risks using counts of fatalities from 1998 to 2000 drawn from the CFOI and hours worked estimates from the CPS. Non-fatal relative risks were calculated using estimated numbers of injury and illness cases with days away from work. These estimates were based on the microdata of the SOII.

In the results section we show that Hispanic men and women tend to work in jobs that are at higher risk of workplace injury and illness. To assess the extent to which the aggregate relative risks for Hispanic men and women are influenced by the distribution of employment, we calculated relative risk measures that standardize for the distribution of employment at the major occupational group level. To standardize at this level we calculated the distribution of hours worked for all employees over the major occupational groups. We then calculated a standardized relative risk for each gender and race/ethnicity group as the weighted average of each group's major occupation relative risks, where the weights are the hours worked shares for all workers. This measure answers the following question: What would be the relative risk for a group of workers if the hours worked by those workers were distributed in the same way as all workers but where their fatality or injury rates were those they actually face?

Some data limitations affected the measures of relative risk and the scope of the non-fatal measures. Both the CFOI and the SOII suffer from varying degrees of item non-response. Data are sometimes missing for occupation, gender, race/ethnicity, and industry.

There are two ways to address item non-response in calculating relative risk measures: Exclude all cases with missing data or impute values when they are missing. By excluding all cases with missing data the implicit assumption is that the missing data would exactly follow the pattern of observed data (e.g., if whites were 70 percent of the observed data, they would also be 70 percent of the missing data). Omitting cases with missing data would bias downward rates of injury or fatality, since the numerators of the rates would be less than they would be without missing data. But, the level of relative risk is unaffected by omitting cases with missing data, since the relative risk for all workers is standardized to 1.

Imputing for missing data is a more sophisticated approach. It uses other covariate information to infer the missing characteristics. It allows patterns of values for the missing cases to differ from those in the observed data.

In calculating our relative risk measures we dealt differently with the cases with missing values depending on the severity of the problem. Missing data is not a very great problem in the CFOI.[1,2] Further, imputation techniques would be relatively crude since the CFOI data set is relatively small. Hence, we chose to omit cases in CFOI where data were missing.

Missing data are a problem in the non-fatal survey for race/ethnicity[3]. We chose to impute for missing gender and race/ethnicity using a nearest neighbor approach. This technique assigns to each case with missing data a value from a donor case. The case with missing data and the donor case are matched based on shared characteristics. To impute gender we matched missing and donor cases based on detailed occupation, and major industry group. To impute race we used state, gender (actual or imputed), occupation and major industry group. To evaluate the result of the imputation we computed the distribution of cases by gender and race/ethnicity for the cases with only non-missing data and for the cases with missing but imputed data. The distributions were quite similar.

The scope of the relative risk measures is different for fatalities and non-fatal cases. Relative risk measures for fatalities are calculated for all civilian workers. That is, the relative risk for a particular group of workers is the rate of fatalities for all civilian workers in that group relative to the rate of fatalities for all civilian workers.

The relative risk analysis of non-fatal injuries and illnesses is narrower in scope than the analysis of fatalities. The non-fatal survey applies only to private industry workers, so that the self-employed, government workers and private household workers are excluded. The race/ethnicity data in the non-fatal survey are always missing for the following four industries: metal mining (SIC 10), coal mining (SIC 12), nonmetallic minerals (SIC 14), and railroad transportation (SIC 40)[4]. We did not feel that it was appropriate to impute race/ethnicity to these industries since we had no information about the race/ethnicity of injured workers. Thus, we excluded these from the analysis. Finally, the non-fatal survey does not obtain injury data for farms with fewer than 11 employees. In order to calculate relative risks it is necessary to align the scope of the injury and hours worked data but the CPS does not allow us to identify workers on small farms. We were forced to, therefore, exclude workers in agricultural production (SICs 01 and 02). Thus, the relative risks for non-fatal cases are for private industry excluding SICs 01, 02, 10, 12, 14, and 40.

---

[1] In the 1998-2000 CFOI, data are missing in the following proportions for key variables: industry 0.5 percent, occupation 0.6 percent, race/Hispanic 1.4 percent.

[2] For example, for calendar year 2000, on a weighted basis, race/Hispanic was missing for 28.8 percent of cases. However, as noted below in the text, race/ethnicity is always missing for SICs 10, 12, 14, and 40, so that the percent of cases with missing race is lower when these industries are excluded.

[3] Data for these industries come from sources outside the Bureau of Labor Statistics and these sources do not obtain race and Hispanic ethnicity.

[4] Data for these industries come from sources outside the Bureau of Labor Statistics and these sources do not obtain race and Hispanic ethnicity.

In calculating relative risk measures by industry we had to recode census industry codes that are used in the CPS to Standard Industrial Classification industries that are used in the SOII and CFOI. This is straightforward at the major industry division level, since with one exception, each census industry maps into one of the major industry divisions reported here. The one exception is the census industry "not specified manufacturing industries," which maps into neither durable or nondurable manufacturing. We mapped the few CPS observations for this industry into the SIC industry "miscellaneous manufacturing industries" (SIC 39), which is part of durable manufacturing.

Both the CPS and the CFOI have separate variables for race and for Hispanic ethnicity. We combined these variables to generate a single race/ethnicity variable indicating whether a worker or decedent was white non-Hispanic, black non-Hispanic, Hispanic, and other non-Hispanic. A white non-Hispanic value was created for the race/ethnicity variable if the race code was white and the Hispanic code was non-Hispanic. Similarly, black non-Hispanic and other non-Hispanic values were generated if the race code indicated black or another race (but not missing), respectively, and the Hispanic code indicated non-Hispanic. Finally, if the Hispanic variable indicated Hispanic, a value for Hispanic was generated for the race/ethnicity variable regardless of the race code (including missing race). If the value for race was missing and the worker was non-Hispanic or the value for Hispanic was missing, then the worker received a missing value for the race/ethnicity variable.

Appendix D-2

Additional Data Sources

Many databases, from income and expenditures to health and educational status, exist that provide informative data on the characteristics of Hispanics in the United States. However, there are few programs other than the CFOI the SOII, and the National Traumatic Occupational Fatality program that provide detailed information on occupational injuries to Hispanic workers. Some of the other databases that do provide such information are listed below. Also listed are databases that collect data on Hispanic workers, although not necessarily related to occupational health. This list below is not exhaustive but may provide a helpful starting point for researchers looking for additional data sources.

- Census of Fatal Occupational Injuries

The Census of Fatal Occupational Injuries (CFOI) collects detailed information on all work-related fatalities resulting from injury during a given year, including demographic data such as employee work status, gender, age, race or ethnic origin; occupation information; employment industry; the event or exposure causing the fatality; the sources of the injury; the activity of the worker during the time of the incident; and the location in which the fatal injury took place.
For more information see <http://www.bls.gov/iif/oshcfoi1.htm>.

- Survey of Occupational Injuries and Illnesses

The Survey of Occupational Injuries and Illnesses (SOII) is a mandatory survey that collects data on non-fatal workplace injuries and illnesses from a stratified random sample of approximately 176,000 private industry establishments. Excluded from the data collection are government agencies, farms with fewer than 11 employees, and the self-employed. The following elements are collected: gender, age, race/ethnicity, and occupation of the worker, as well as the nature of injury (e.g., sprain, fracture), part of body, source of injury or illness (e.g., tool, surface), and the event (e.g., fall, assault).
Case and demographics: http://www.bls.gov/iif/oshcdnew.htm
Incidence rates: http://www.bls.gov/iif/oshsum.htm

- National Traumatic Occupational Fatality Surveillance System

The National Traumatic Occupational Fatality (NTOF) database includes only data from death certificates indicating that (1) death was related to external causes; (2) the deceased was greater than or equal to 16 years of age at the time of death; and (3) the injury occurred at work.

- Fatality Assessment and Control Evaluation

The Fatality Assessment and Control Evaluation (FACE) surveillance system contains first reports of traumatic occupational fatalities in 15 states obtained through multiple sources of notification including death certificates, coroner and medical examiner reports, OSHA, law enforcement, the media, and other injury surveillance systems. Currently, FACE first reports are limited to 15 states. Participating states change from year to year because states enter and leave the program based on competition for funding.

- National Agricultural Workers Survey (NAWS) Youth Injury and Farm Worker Health Supplement

The survey will collect information on farm-related injuries occurring to workers under 20 years of age. Ethnicity categories include Mexican, Puerto Rican, Central/South American, other Hispanic, non-Hispanic, and unknown.

- Common Information Service System (CISS)— Mining

CISS is a mining information system provided by NIOSH Mining Safety and Health Research (formerly the United States Bureau of Mines).

- Occupational Safety and Health Guidelines for Chemical Hazards

Occupational Safety and Health Guidelines for Chemical Hazards summarizes information on permissible exposure limits, chemical and physical properties, and health hazards. It provides recommendations for medical surveillance, respiratory protection, and personal protection and sanitation practices for specific chemicals that have federal occupational safety and health regulations.

- National Occupational Mortality Surveillance System (NOMS)

The purpose of this data system is to provide a resource for surveillance and research in occupational health. The states in the system are not consistent from year to year. The Hispanic data is not complete for all years.
For more information see<http://www.hhs.gov/aspe/minority/mincdc30.htm>.

- National Surveillance of Non-fatal Occupational Injury

In collaboration with the National Institute for Occupational Safety and Health (NIOSH), the CPSC collects injury information on civilian work-related injuries treated in 67 hospital emergency departments. Ethnicity categories may be included in a free-text field for "race = other."

- The Alaska Trauma Registry (ATR) collects data from all hospitalized traumatic injuries that are admitted to any of the 24 hospitals in Alaska. ATR is administered with assistance from NIOSH and focuses on work-related injury surveillance. Information on Hispanic origin is included.

- The Coal Workers' X-Ray Surveillance program provides respiratory health screening and surveillance to monitor trends in coal workers' pneumoconiosis in United States miners. This program contributes surveillance data for prevention activities to reduce the burden of coal workers' pneumoconiosis and related lung diseases.

- Hispanic Population of the United States from the Current Population Survey

The CPS core survey is the primary source of information on the employment characteristics of the civilian non-institutional population, ages 16 and older, including estimates of unemployment released every month by the Bureau of Labor Statistics. This survey includes data on income and poverty levels.
For more information see <http://www.census.gov/population/www/socdemo/hispanic.html>.

- Hispanic Health and Nutrition Examination Survey (HHANES)

This survey contains the following data on Hispanic workers: demographic characteristics, acculturation, cardiovascular conditions, health condition list, diabetes, functional impairment, digestive disease, health services use, hypertension, meal programs, pesticide exposure, smoking, vision, hearing, reproductive health, and selected conditions For more information see<http://www.cdc.gov/nchs/products/catalogs/subject/hhanes.htm >.

- National Agricultural Statistics Services

This database contains information on Hispanic farm characteristics such as number of farms, farms by size, land use, value of commodities, net cash return, tenure, and average age of operator. For more information see <http://www.usda.gov/nass>.

- Consumer Expenditure Survey

This survey provides information on the buying habits of United States consumers, including data on their expenditures, income, and consumer unit (families and single consumers) characteristics (includes Hispanic origin). For more information see <http://www.bls.gov/cex>.

- National Center for Education Statistics

The NCES provides tables and reports on Hispanic Dropout Rates by Immigration Status. For more information see <http://nces.ed.gov/pubs2001/dropout/StatusRates3.asp>.

- Bureau of Justice Statistics
    - Annual Survey of Jails

    Information is available on the number of inmates by sex, race, adult or juvenile status, reason being held, and cause of death. Facility characteristics are collected regarding capacity, court orders, conditions of confinement, alternative programs, and average daily population.
    - Capital Punishment in the United States
    - National Crime Victimization Survey

      Information about the victims of crime in this survey includes race.
    - National Judicial Reporting Program

      Convicted felons by sociodemographic characteristics including race are covered by this program.
    - Survey of Adults on Probation

For more information see<http://www.icpsr.umich.edu/NACJD/bjs.html#cp>.

- Residential Energy Consumption Survey

Information is provided on the use of energy and includes demographic characteristics of the household. For more information see <http://www.eia.doe.gov/emeu/recs/contents.html>.

- American Housing Survey

This survey provides data for evaluating progress made toward "a decent home and a suitable living environment for every American family" and includes demographic, financial, and mobility characteristics of the occupants. For more information see <http://www.census.gov/hhes/www/ahs.html>.

- Continuing Survey of Food Intakes by Individuals

The Continuing Survey of Food Intakes by Individuals (CSFII) is designed to measure what Americans eat and drink (includes demographic characteristics). For more information see <http://www.barc.usda.gov/bhnrc/foodsurvey/home.htm>.

- Early Childhood Longitudinal Study (ECLS) Program

This program measures children's status at birth and at various points thereafter; children's transitions to non-parental care, early education programs, and school; and children's experiences and growth through the fifth grade. For more information see <http://nces.ed.gov/ecls>.

- National Assessment of Educational Progress (NAEP)

The National Assessment of Educational Progress (NAEP) is mandated by Congress to monitor the knowledge, skills, and performance of the nation's children and youth (includes demographic characteristics). For more information see <http://www.ed.gov/NCES/naep>.

- National Health Interview Survey

This survey consists of data about illnesses, injuries, impairments, chronic conditions, activity limitation caused by chronic conditions, use of health services, and other health topics. The study was designed to allow the development of national estimates of health conditions, health service use, and health problems of the United States civilian non-institutionalized population. For more information see <http://www.cdc.gov/nchs/nhis.htm>.

- National Immunization Survey

Information is collected on the vaccinations received by children 19-35 months old. For more information see <http://www.nisabt.org>.

- National Vital Statistics System

This system collects and publishes data on births and deaths in the United States. For more information see <http://www.cdc.gov/nchs/nvss.htm>.

- Population Projections

Produces projections of the resident population for the nation and for each of the 50 states and the District of Columbia. Projections of the number of future households and families in the United States are also produced at the national level. For more information see <http://www.census.gov/population/www/projections/popproj.html>.

- Survey of Income and Program Participation

The SIPP collects detailed information on income, labor force participation, participation in government assistance programs, and general demographic characteristics to measure the effectiveness of existing government programs, to estimate future costs and coverage of government programs, and to provide statistics on the distribution of income in the United States. In addition, topical modules provide detailed information on a variety of subjects, including health insurance, child care, adult and child well-being, marital and fertility history, and education and training. The United States Census Bureau releases cross-sectional, topical modules and longitudinal reports and data files. For more information see <http://www.bls.census.gov/sipp>.

- Uniform Crime Reports

These reports provide information on the following crimes reported to law enforcement authorities: homicide, forcible rape, robbery, aggravated assault, burglary, larceny-theft, motor vehicle theft, and arson. For more information see <http://www.fbi.gov>.

- Immigration and Naturalization Service (INS)

The *Statistical Yearbook* provides information about the various types of foreign nationals who are inspected, naturalized, apprehended, or removed by the INS. Types of aliens include immigrants, non-immigrants (temporary visitors), parolees, refugees, and those seeking asylum, as well as those naturalized or apprehended. For more information see <http://www.ins.gov/graphics/aboutins/statistics/ybpage.htm>.

References

Camarota, Steven A. 2001. Immigrants in the United States-2000. A Snapshot of America's Foreign Born. : Center for Immigration Studies National Council of La Raza. 2001.Beyond the Census: Hispanics and the American Agenda. Report. 8/01. 24 October 2002 <http://www.cis.org/articles/2001/back101.html>

Camarota, Steven A. 1997. 5 million illegal immigrants. Center for Immigration Studies. Immigration Review 28 (Spring): , 24 October 2002 <http://www.nclr.org/policy/census/census_report01_part_I.pdf>

Economist 2001. Between Here and There. 7/7/01. London, The Economist Newspaper, Ltd.

National Council of La Raza. 2001.Beyond the Census: Hispanics and the American Agenda. Report. 24 October 2002 <http://www.nclr.org/policy/census/census_report01_part_I.pdf>

U.S. Department of Commerce, Census Bureau. 2001 [1]. The Hispanic Population. Census 2000 Brief. Washington, D.C. 24 October 2002 <http://www.census.gov/prod/2001pubs/c2kbr01-3.pdf>

U.S. Department of Commerce, Census Bureau. 2001 [2]. Census 2000 Paints Statistical Portrait of the Nation's Hispanic Population. News Release. Washington, D.C. 24 October 2002 <http://www.census.gov/Press-Release/www/2001/cb01-81.html>

U.S. Department of Commerce, Census Bureau. 2001 [3]. Projections of the Total Resident Population by 5-Year Age Groups, etc. Population Projection Program #NT-P4-G.

U.S. Department of Commerce, Census Bureau. 2001 [4]. Population by Race and Hispanic or Latino Origin for the U.S., etc. Table 5 #PHC-T-6. Washington, DC. 24 October 2002 <http://www.nclr.org/policy/census/census_report01_part_I.pdf>

U.S. Department of Commerce, Census Bureau. 2001 [5]. Profile of the Foreign-Born Population in the United States: 2000. Special Study Report. 24 October 2002 <http://www.census.gov/Press-Release/www/2000/cb00-164.html>.

U.S. Department of Commerce, Census Bureau. 2000. Census 2000: Hispanics in the U.S. Presentation. 24 October 2002 <http://www.census.gov/mso/www/rsf/hisorig/sld001.htm>.

Appendix E

An Examination of Occupational Safety and Health Materials Currently Available in Spanish for Workers as of 1999

Marianne P. Brown, Director
UCLA-LOSH Program

INTRODUCTION AND STATEMENT OF THE PROBLEM

It is becoming increasing clear that Spanish-speaking workers in the United States work in some of the most dangerous industries, and they currently have the highest occupational fatality rate of any ethnic group. While workplace deaths for white and black workers declined in 2000, deaths for Hispanic/Latino workers increased sharply from 729 in 1999 to 815 in 2000, with a 24 percent jump in construction fatalities (Hedges, 2001). The extent of the problem is not necessarily captured in this Bureau of Labor Statistics data for a variety of reasons. A Newsday investigative series in July 2001 reported that "OSHA (Occupational Health and Safety Adminstration) officials say immigrant deaths in construction and manufacturing are far more likely than retail or agriculture to come to their attention because they are often unionized jobs where safety violations are more frequently reported" (Maier, 2001). In Newsday's study of New York state's workplace deaths in a six-year period, deaths were found to occur "in retail places such as late-night restaurants, gas stations, and other small, cash-only businesses, yet only a small percentage of all OSHA fatality investigations occurred in this area" (Maier, 2001). In addition to possible under-reporting of workplace fatalities, it is likely there is also under-reporting of workplace injuries and illnesses by Spanish-speaking workers and their employers, as these can be "hidden" more easily than a fatality.

The reasons for such non-reporting are many, but a major factor is that many of these workers are undocumented, without legal work papers. It has been estimated that in some industries, particularly in California, undocumented workers account for 50 percent or more of the workforce (Cleeland, 2000). The current head of federal OSHA, John L. Henshaw, has acknowledged that "we recognize that employers who hire undocumented workers may be afraid to report workplace deaths, due to possible legal repercussions from their hiring practices" (Henshaw, 2002). Undocumented workers particularly are often afraid to report injuries or illnesses for fear they will be fired or turned in to the Immigration and Naturalization Service by the employers.

According to the National Institute for Occupational Safety and Health (NIOSH), "The number of Hispanic workers in the U.S. work force is expected to increase by more than one-third over the next decade" (NIOSH, 2001). This fact, combined with the disproportionate number of Hispanic/Latino workplace fatalities in 2000, 13.8 percent, vis-à-vis their proportion of employment, 10.7 percent, points to the need for new interventions to stop this rising tide.

(It should also be noted that it is not only in California, New York, New Jersey, and Florida where a significant number of Spanish-speaking people are now working. There are significant numbers in North Carolina, Georgia, Illinois, Nebraska, Maine, and other states as industries in these states recruit low-wage immigrant workers to improve their profit margins.)

Part of the solution is to provide information and training to these workers in their first language in an accurate and culturally sensitive manner. Health and safety professionals point out that "Hispanic immigrants, partly because many do not speak English, often receive less job and safety training than American-born workers do. Language barriers often contribute to the higher

Hispanic injury rate .... Many job sites, safety instructions and warnings appear only in English" (Greenhouse, 2001).

It is important to emphasize, however, that information and training cannot be the only preventive action taken. Putting all the burden on the worker to "work safely" reflects a limited understanding of how to make the workplace safe and also is unfair to the worker. Engineering controls is the most desirable and effective safety method (e.g., putting guards on cutting machines, having a lockout system in place when doing electrical maintenance). Needed personal protective equipment must also be an integral part of the safety program (e.g., providing harnesses and safety belts when doing elevated work). Providing eating and drinking areas that are not contaminated with lead or other workplace toxics is another necessity. Once these control measures are in place, information and training for the worker will cover why the guards are there and must stay in place, steps to follow when locking out a machine, why and how to wear safety harnesses, and why and how to practice good hygiene so that one's food or drink is not contaminated with workplace toxics.

METHODOLOGY

To write this paper I conducted a review of over 500 educational health and safety materials developed by over 75 organizations for workers up to 1999. I have also examined approximately 50 such materials developed since 1999. All were primarily written materials (booklets, pamphlets, fact sheets) and posters that were produced by governmental agencies at the federal, state, and local levels; university programs; unions; worker advocacy groups; occupational health and safety professional associations; voluntary health agencies; for-profit companies that produce health and safety materials; and industry. In addition, at least a fourth of the materials reviewed were from Spain and Latin American countries. Between 1997 and1999, the program that I direct, the UCLA-Labor Occupational Safety and Health (LOSH) program updated a bibliography of such materials that we originally produced in 1990, La Fuente Obrera: Materiales en Español de Salud y Seguridad Para Trabajadores y Profesionales de Salud; this Spanish/English bibliography is available on our website at <http:// www.losh.ucla.edu> (Alas, 1999). Since that time, we also have acquired new material. We have at least one copy of each of the materials reviewed in our Spanish Resource Library. People can request from us photocopies of materials that do not have copyright restrictions.

FINDINGS

In this updated review as of Spring 2002 it has again become clear that although there have been numerous materials developed for Spanish-speaking workers, there are still many gaps. There are needs for both new materials and revisions of those that exist. Here is a snapshot of what we found.

Federal OSHA, under the directorship of Dr. Eula Bingham from 1976 to 1980, produced a series of Spanish-language booklets, a manual/workbook, and a poster, which are now out of print. These were compact, in lively colors with graphics, and covered topics such as "Health and Safety Committees: A Good Way to Protect Workers"; "Safety and Health at Work: Answers to Some Common Questions"; "Safety and Health at Work: OSHA Inspections from Start to Finish"; a manual/workbook for workers on "Health Inspections from OSHA: How You are Able to Help"; and a poster entitled "They Cannot Punish You for Insisting on Safety and Health on the Job. It is the Law". Currently OSHA has four booklets on "Risks of Chemicals," "Regulations," "Bloodborne Pathogens," and "Worker Rights" that were developed in the period from 1988 to 1992; they are text-heavy without many graphics. OSHA also has three information cards about heat stress, cold stress, and damaging rays from the sun, and one poster. None of OSHA's videos for loan are in Spanish.

NIOSH in 1982 partnered with the Instituto Nacional de Seguridad e Higiene en el Trabajo in Barcelona, Spain, and translated NIOSH's Pocket Guide to Chemical Hazards into Spanish. Between 1984 and 1986 NIOSH created 10 NIOSH Alertas of about four pages each that describe cases in which workers, for example, died in confined spaces or were electrocuted doing certain kinds of work and then give recommendations for how these kinds of deaths can be avoided in the future. These Alertas are dense with text and have no illustrations. Currently one can download 26 Spanish publications from their website, but these are text-heavy and often lengthy. One brochure "Eres Un Joven Que Trabaja?" intended for teen workers was originally produced by UC Berkeley with NIOSH funds and then was put on the NIOSH website with all but one illustration removed, resulting in a much less attractive and user-friendly item. NIOSH has funded two attractive, colorful, highly graphic brochures on skin cancer that were produced by the Telamon Corporation and PATH: "Corre Peligro Trabajando Bajo el Sol?" and "Protegete del Sol." Through NIOSH's Fatality Assessment and Control Evaluation (FACE) program, 2 of the 15 programs have four fact sheets in Spanish on their websites. One program has its Case Reports and one Hazard Alert in Spanish.

The Mine Safety and Health Administration does not have anything on their website in Spanish, but they have produced eight videos in Spanish – these are dubbed -- with more to be finished by the end of this summer. They also have a number of booklets and information cards in Spanish on topics such as silicosis, mercury, machine guards, personal protective equipment, etc. Some, not most of the materials use graphics to illustrate points. They do appear to have developed more materials for workers in Spanish than has OSHA.

The Centers for Disease Control and Prevention has a "CDC en Español" section on their website. Within that is a section on salud occupacional that links to some NIOSH publications in Spanish. It is surprising that the CDC Information Networks page does not have a link to NIOSH.

The Environmental Protection Agency has Spanish training material for workers available through their Office of Pesticide Programs (OPP) which has a Spanish translated version of the OPP homepage. It is not stated on the homepage that they have these materials, however.

The **New Jersey Department of Health Services** translated into Spanish all the fact sheets they had developed in English between 1986 and 1989. These are more user-friendly than most safety data sheets and can be used as references when doing worker training. These should be available in workplaces where there are Spanish-speaking workers in order to meet the intent of the workers' right to know standards.

California's Department of Health Services' Occupational Health Branch has developed a number of educational materials in Spanish, including fact sheets on formaldehyde, methyl bromide, methylene chloride, pregnancy and working with chemicals, pulmonary function tests, and mold. They also have a poster and Guide to Solvent Safety, a booklet on ergonomic issues for sewing machine operators and for jewelry workers (including a poster), and a trainee manual for workers (excluding construction workers) who work around lead, "Como Prevenir el Envenenamiento con Plomo en el Trabajo." All these publications are California-specific, so they would need to be rewritten for use elsewhere.

California-OSHA's Consultation Office has developed some materials in Spanish, mostly for the agricultural sector: a guidebook for farm labor contractors, an informational book on Cal-OSHA's special emphasis program in agriculture (ASHIP), and a poster targeted at farm labor contractors saying what they must provide for the workers. For employers Cal-OSHA's Construction Safety and Health Inspection project has a poster that specifies which areas in the workplace will be targeted for inspections.

University labor education programs, such as the UCLA Labor Occupational Safety and Health (UCLA-LOSH) program and UC Berkeley's Labor Occupational Health Program (LOHP), have tried to incorporate in their materials good principles of how adults learn best and how to develop materials for workers with limited literacy (Szudy, 1994). Some of their products include LOSH's packet of 11 Spanish/English fact sheets that includes information on workers' rights to bathroom breaks, how to identify hazards in the workplace, Employer Log 300[1] requirements, ergonomics, and workers' health and safety rights. LOHP has produced a bilingual photonovela entitled "Le Enferma Su Trabajo?" that has been used both in the United States and in Mexico.

Since federal OSHA's New Directions in Worker Training funding in the late 1970s, unions have produced a variety of brochures, fact sheets, manuals, and posters in Spanish. There is a lot of variability in quality and user-friendliness in Spanish translation of these union publications. Compared with the government publications, the union publications tend to have more graphics that are not as "serious" as the governmental ones (they are more likely to be in a cartoon style). The unions that have been most active in this area are:

- Union of Needletrades,
- Industrial and Textile Employees (UNITE)
- International Ladies Garment Workers Union (ILGWU)
- Amalgamated Clothing and Textile Workers Union (ACTU) (The ILGWU and the ACTU came together to form UNITE about five years ago.)
- International Union of United Automobile
- Aerospace and Agricultural Implement Workers of America (UAW)
- Canadian Auto Workers Union
- Retail, Wholesale and Department Store Union (RWDSU)
- United Food and Commercial Workers (UFCW)
- Service Employees International Union (SEIU)
- American Federation of State, County and Municipal Employees (AFSCME)
- Laborers International Union of North America (LIUNA)
- Glass, Pottery, Plastics and Allied Workers Union (GPPAWU).

The topics most often covered are how to identify hazards, workers' health and safety rights, health and safety committees, and specific hazards. There have been collaborations between governmental organizations and unions, such as with EPA's booklet "Asbestos en Edificios: Guia para el Personal de Servicio y Mantenimiento," which was adapted by the Service Employees International Union (SEIU) with some changes. For one thing SEIU added a section on the limitations of respirators and also alerted the service and maintenance workers that if they worked in a school, there is another law (AHERA) to protect them from asbestos exposure.

Worker advocacy groups also have developed Spanish-language materials on health and safety. Such groups include: the Center to Protect Workers' Rights in Washington, D.C., which has created Hazard Alerts on construction topics, which are available in pocket-size cards or can be downloaded from their website; the Committees on Occupational Safety and Health (COSH groups)—there are about 25 of these largely volunteer groups around the United States and they have developed mostly fact sheets of one to two pages that are usually illustrated; and the Coalition for Justice in the Maquiladoras in San Antonio, Texas, which has developed comic-book-style publications that answer, "Cuáles son mis Derechos en el Trabajo?"

---

[1] The employer is required under OSHA law to post a summary log every February of injury and illnesses that have been reported for the past year in the work place.

Voluntary health agencies such as the American Lung Association have developed written and audiovisual materials on workplace health and safety. The branches in Chicago and Los Angeles developed some materials in the 1980s.

Latino/Hispanic Advocacy Organizations do not appear to have much available on workplace health and safety for workers. The National Council of La Raza, for example, has nothing on this topic on their website. The National Alliance for Hispanic Health does have some fact sheets on asbestos, dusts and debris (related to the September 11th, 2001 cleanup in New York City) and anthrax. These only have text, no illustrations.

Occupational safety and health professional organizations such as the American Industrial Hygiene Association have developed some materials. A 1995 booklet entitled "Un Enfoque Ergonómico para Evitar Lesiones en el Lugar de Trabajo" has no graphics or illustrations and is very dense with text. A 121-page book Protección Respiratoria: Un Manual y Guía (second edition printed in 1991) has been written for the persons responsible for initiating and operating respiratory protection programs in the workplace. It is very technical with some black-and-white photographs.

Employer Organizations have created some training manuals for workers in Spanish, such as the Association of General Contractors. But no such materials are listed on their website. The National Safety Council's Safety Center, Inc. has developed Spanish-language materials for agricultural workers which can be downloaded from AgSafe's website. These are short fact sheets with no graphics.

Commercial health and safety companies have also developed some materials that are of varying quality and usefulness. For example, Genium Publishing Corporation in 1990 created "El Diccionario de Bolsillo de las MSDS." This guide describes the terms and concepts commonly used in material safety data sheets . In 1995 Genium started publishing material safety data sheets in Spanish; the sheets were prepared between 1985 and 1995 and come in two three-ring binders. Krames Communications has a number of comic-book-type publications on occupational safety and health topics, such as back injuries and hazard communication. Krames publications tend to put all the responsibility for safety on the shoulders of the workers, and sometimes the illustrations could be offensive to the reader. Coastal Video Communications Corporation in Virginia Beach, Virginia, has produced booklets on asbestos, lockout, confined spaces, and other topics that are accompanied by videos on the same topics. They are in color, have many drawings, and are generally of good quality. The Dow Chemical Company has developed a pocket-size booklet "Como 'llevarse bien' con los Solventes," originally prepared in 1975 for NIOSH. It is in a very simple format with line drawings on every page to illustrate the main point. It emphasizes that working safely with solvents is less dangerous than driving a car in heavy traffic.

Our samples of educational materials from sources outside the United States ran the gamut from being very technical and dense with text, to being very worker-friendly with many graphics, cartoons, and photographs. Many of the materials were in a small paperback format and newsprint was often used for the inside.

- The Center for the Study of Workers' Health in Ecuador, for example, has a series of such booklets on such topics as "Evaluación de los Riesgos por un Grupo de Trabajadores" (making risk maps is part of this) to "Safety on Machines."
- The International Federation of Industrial Workers in Belgium (ICEF) has a guide for working with solvents that has good illustrations and a centerfold showing the proper gloves to wear with different kinds of solvents.
- Nicaragua's Programa Centroamericano de Ciencias de la Salud del CSUCA produced a 90-page Manual Práctico de Seguridad Industrial under the direction of the Minister of Work that is done in a comic-book style with unsophisticated but appealing line drawings.

- In Madrid, Spain, the Instituto Nacional de Seguridad e Higiene en el Trabajo and the Unión General de Trabajadores have created eight booklets on such topics as "El Plomo Y Nuestra Salud" that are written simply with often humorous illustrations.
- El Instituto Salud y Trabajo in Peru has a series on risk factors, such as Ruido, that are bi-fold brochures with little text and humorous illustrations. The institute has also collaborated with two other organizations there to produce a bi-fold brochure about "Silicosis y Otros Riesgos en el Trabajo de Producción de Refractarios" (heat resistant bricks).
- Peru's Centro de Información Estudios y Documentación has developed a series of booklets on health and safety; one such is "Los Textiles," which in a simple comic-book style, talks about the safety, health risks (including noise, chemical, biological, and sanitation) in the textile manufacturing industry.
- The Instituto Mexicano para el Desarrollo Comunitario in Guadalajara, Mexico, developed a questionnaire in a light-hearted, comic-book format that asks the workers what they thought of the health and safety materials they developed (a list was attached) and how useful were they to the workers.

The International Labor Organization's Occupational Safety and Health Branch has a thirteen module curriculum—the modules come in a cardboard carrying case—"Su Salud y Seguridad en el Trabajo" that is available both in hard copy and in CD Rom. They also have a manual "Lista de Comprobacion: Ergonómica" which has graphics and isuser-friendly.

In summary, there is a lot of Spanish-language material that has been developed for workers on health and safety. The quality is mixed, at best. Most of it is not useful for workers who have limited literacy and little education. Instead, the materials are too technical, too wordy, with little to no graphics to illustrate the key points.

RECOMMENDATIONS

The Need for a Systematic Approach to Materials Development

From this review it appears that a fair amount of Spanish-language educational materials have been developed for both the construction and agricultural sectors, where Spanish-speaking workers have long worked in the United States. These have been developed for the construction trades primarily by commercial vendors and by the Center to Protect Worker Rights of the Building and Construction Trades Department of the AFL-CIO, which has developed 20 pocket cards (and flyers when the Spanish version was too long to fit on a pocket card). For the agricultural sector these have been developed primarily by the United Farm Workers Union, some coalitions on occupational safety and health (COSH) groups, the National Safety Council, The Environmental Protection Agency, and state health agencies. Although these materials exist, there is a need to make them user-friendlier with graphics and less dense text.

Most importantly, at this point there is a need to target the other industries where Spanish-speaking workers predominate to assess where the gaps are regarding available, good quality, Spanish-language materials for those industries and to identify organizations that can develop those materials. Such industries include the manufacturing sector: garment work, textile work, lead battery plants, food processing and packing, furniture manufacturing, and electronics. In the service sector Spanish-speaking workers toil in large numbers: healthcare and homecare, domestic work, gardening, airport services, parking lot attendants, hotel, restaurant, auto repair, and janitorial and maintenance workers, to name a few.

With respect to janitorial and maintenance workers, there is now a need for Spanish-language information for those who clean up after bioterrorist attacks. This became apparent when it was found that many of those who cleaned office buildings near Ground Zero in New York City were immigrant workers from Colombia and Ecuador. In fact, of the 415 such workers examined by a mobile health unit at the site all the workers had health symptoms. They did not receive any health and safety information or training prior to doing the work. Instead, they were "given mops, rags and bags and told to remove inches of dust that coated the floors, walls and desks in offices. Most said they were not given protective equipment. Some workers who brought their own respirators said employers told them not to wear such protection" (Ramirez, 2002).

In addition to sector-specific or job-specific materials, there is a need for educational materials on worker health and safety rights, including those that address the issue of rights of undocumented workers. Up until the March 27, 2002, Supreme Court ruling that an undocumented worker could not collect back pay after he was illegally fired for union-organizing activities, the Department of Labor has said that wage, hour, and safety regulations would be enforced vigorously for all workers, regardless of legal status. Since this ruling, employers have attempted, albeit it unsuccessfully, to fire workers when they raised health and safety issues. There is now a widespread belief among illegal immigrants that they no longer are protected under state and federal labor laws.

As well as learning their rights, Spanish-speaking workers need to know where they can go for information about the hazards in their workplaces and where to turn to inside or outside their workplaces for additional information or assistance. The referrals, of course, will need to be to places where Spanish-language materials and Spanish-speaking individuals can assist.

How The Materials Are Developed is Also Important

There are a number of guidelines to follow when developing Spanish-language materials: It is undesirable to do straight translations from English materials, have someone who speaks Spanish on a daily basis write the text, keep the text at a medium-to-low literacy level, use plenty of graphics, and pilot test the material with a subgroup of the kinds of workers for whom it is intended so that you can get their input on content and mode of presentation.

It is better to write the material in Spanish to begin with, because then there will be less of a risk that the writer is trying to conform to an English way of saying things. The writer then will be "freer" to say things in a more culturally sensitive and, hopefully, less technical way. If something is written in English first and then translated into Spanish, it will probably be a longer publication, because it often takes more words to say the same thing in Spanish. It is advisable to put key words in English also, for example, *montacargas*/forklifts. There is another reason to do this also; certain English words in particular, such as "Danger," "Hard Hat Area," "Do Not Enter," "OSHA," need to be understood by non-English speakers. It is always good to have an English version of whatever you have in Spanish in case the supervisor or co-workers speak English as their first language. There is a very effective video created by the California Department of Health Services' Occupational Health Branch, "He's Not the Man I Married ...." (OHB) about lead exposure on the job that has part of the dialogue in Spanish and part in English. It is done in such an artful way that those who view it who are either English-speakers or Spanish-speakers can both understand what is taking place.

Another thing to remember is that not all Spanish-speaking people have the same Spanish words for things. For example, in Nicaragua the common word for forklifts is *mulas*, but in Mexico it is *montacargas*. Puerto Rican Spanish is different from Mexican, and Mexican is different from El Salvadoran. Therefore, whenever possible, the translation should be in as generic a Spanish as possible.

Have someone who speaks Spanish on a daily basis, preferably a native speaker, write the Spanish text. Ideally, it should be someone who knows the topic well also. The computerized translation systems and computerized dictionaries can have errors or they may not capture the needed subtleties. Even comprehensive, well-respected dictionaries can lead you down the wrong path. For example, when looking up the Spanish word for forklifts, in Larousse's Gran Diccionario it says that the correct word to use is *carretilla* or *elevadora*, and that *montacargas* means freight elevator. But, Spanish-speaking workers in California, at least, use *montacargas* as the term for forklifts. It is interesting to note that the NIOSH FACE program in Nebraska uses the term *carretilla elevadora hidráulica* on their fact sheet. Either this is what Spanish-speaking workers there understand to be forklifts or it is an example of a misleading translation.

We have learned also that some words that are not so " loaded" in English can be very "loaded" in Spanish, and we need to be sensitive to these cultural differences. An example of this occurred when we were conducting ergonomic training; we found that workers were less likely to respond that they had pains (*dolores*), but they were willing to say that they had discomforts (*molestias*).

It is important to have a native speaker help choose the graphics that are used so that the publication is both culturally sensitive and appealing to the intended readership. It is important to note again that the materials developed for workers in Spanish-speaking countries often used graphics that were somewhat humorous while still treating the worker with respect. Perhaps there is a lesson to learn here, as similar publications in the United States tend to be extremely serious.

For the particular Spanish-speaking workers in the United States that need health and safety materials in manufacturing, agricultural, service, retail, and other such industries, it is important to keep the literacy level at medium to low. There are, of course, workers who read at a high level, but in order to reach all workers it is important to keep materials at a limited literacy level. I am not aware of any program that can be used to assess the literacy level of Spanish language materials. Our Labor Occupational Safety and Health (LOSH) program at UCLA developed a simple checklist for assessing the literacy level of such materials (see Box B-1).

For easy readability and in order not to overwhelm the reader, it is important to develop these materials using many graphics, photos, or drawings to illustrate the points made. Another approach is to use a comic book or *fotonovela* style to present the information. Illustrated stories using a comic book format is a popular form of reading material for Spanish-speaking people who do not have advanced reading skills. *Fotonovelas* that have photographs of people and presented in a comic-book style are also well received.

In order to ensure that the material you are developing is going to be useful to workers, it is important to pilot test drafts of materials with a subset of the group for which it is intended. This works best using a focus group of such workers. You can send the draft material to them prior to the group meeting or introduce it at the meeting. Create a list of questions you want to ask about usefulness, clarity, appropriate language and graphics, and other ideas they have on how to make it more useful and appealing to workers so that workers will want to read it. It is best not to have the developers of the material present in order to reduce the possibility that the workers will feel compelled to say something nice about the materials.

A crucial question is how will the materials get into the hands of those who need to know? It is important that the producers of the materials create partnerships with relevant trade associations, worker organizations such as unions and worker advocacy groups, churches, medical providers who serve Spanish-speaking populations, businesses where Spanish-speaking people shop, libraries, and schools their children attend. The Internet at this time is one of the least likely channels to reach those who need to know in this particular target group, although the Internet should not be discounted entirely. Materials could be sent by e-mail to the conduit organizations identified here; such organizations could download the materials and distribute them to their constituencies, members, or clients.

And the ultimate question is how can the availability of these materials assist Spanish-speaking workers in improving health and safety conditions in the workplace? Spanish-speaking workers, whether undocumented or not, are often marginalized in their workplace or their union. In a recent study it was found that trainees, in a hazardous waste worker training program, for whom English was not their main language (mostly Spanish speakers) attempted action for workplace safety improvements as often as English speakers. However, the odds of their correcting problems were half that of the English-speaking workers. It was posited by the authors that, "this was due to a perceived lack of control over organizational resources for change, not simply due to communication barriers" (Cole, 1996).

Box B-1
Checklist for Evaluating Literacy Level

Title of Publication:

Producer's Name/Address:

	Yes	No	N/A
A. Content			
1. Are technical words explained?	☐	☐	☐
2. Are simpler words used where possible?	☐	☐	☐
3. Can you easily identify the key message in the text?	☐	☐	☐
B. Design			
1. Is a large enough typeface used for the main text?	☐	☐	☐
2. Is a serif typeface (with feet) used for the main text?	☐	☐	☐
3. Do your sentences avoid large sections of CAPITAL LETTERS, **bold type**, or *italics*?	☐	☐	☐
C. Organization			
1. Is the key "take home" information easy to find?	☐	☐	☐
2. Are headings and subheadings used to help organize the text?	☐	☐	☐
3. Are the margins wide to allow plenty of white space?	☐	☐	☐
D. Illustrations			
1. Do illustrations help explain the text?	☐	☐	☐
2. Are the illustrations clear and realistic?	☐	☐	☐

Key:

11-10	Low literacy
9-8	Low to medium
7-6	Medium
5-4	Medium to high
3-1	High

SUMMARY

There is a great need for more and better Spanish-language workplace health and safety materials for workers in the United States. There has been a need for some time, but the more recent workplace fatality reports have now brought the Spanish-speaking worker into the spotlight.

A systematic approach should be undertaken that looks at where such workers work, what the risks are, what kinds of materials should be developed, and how they should be developed. Lessons can be learned from what has already been produced, but it is crucial that Spanish-speaking workers be involved in the development of new materials.

ACKNOWLEDGMENTS

The author would like to give a special thanks to the following people for reviewing an earlier draft of this paper and giving me thoughtful comments: Juan DeLara, UCLA Labor Center for Research and Education; Matina Ramirez-Aguilar, Instituto Nacional de Salud Publica in Cuernevaca, Mexico; and Tom O'Connor, Coordinator of the National Committees on Occupational Safety and Health (COSH) Network.

REFERENCES

Alas, Sonia, La Fuente Obrera: Materiales en Español de Salud y Seguridad Para Trabajadores y Profesionales de Salud, University of California at Los Angeles' (UCLA) Labor Occupational Safety and Health (LOSH) Program, 1999.

Cleeland, Nancy, "AFL-CIO Calls for Amnesty for Illegal U.S. Workers," Los Angeles Times, February 17, 2000.

Cole, Brian L. and Marianne Parker Brown, "Action on Worksite Health and Safety Problems: A Follow-up Survey of Workers Participating in a Hazardous Waste Worker Training Program, American Journal of Industrial Medicine, 30: 730-743, (1996).

Greenhouse, Steven, "Hispanic Workers Die at Higher Rate," New York Times, July 16, 2001.

Hedges, Joyce, "Fatalities Among Hispanic Workers Rising; Employers Urged to Provide More Training," Occupational Safety & Health Reporter, The Bureau of National Affairs, Inc., Washington, D.C., September 27, 2001.

Henshaw, John L., Assistant Secretary of Labor for Occupational Safety and Health. Statement before the Subcommittee on Employment, Safety and Training, Committee on Health, Education, Labor and Pensions, United States Senate, February 27, 2002, OSHA Website.

Maier, Thomas, "Forgotten Victims: Agency Rarely Investigates On-the-job Deaths of Immigrants," Newsday, July 23, 2001.

NIOSH Update, NIIOSH Website, December 19, 2001.

OHB (Occupational Health Branch), California Department of Health and Human Services, 1515 Clay Street, Suite 1901, Oakland, California, 94612, 510-622-4300.

Ramirez, Margaret, "World Trade Center Minority Workers' Ills Persist," Newsday, April 28, 2002.

Szudy, Elizabeth and Michele Gonzalez Arroyo, The Right to Understand: Linking Literacy to Health and Safety Training, Labor Occupational Health Program, University of California at Berkeley's Labor Occupational Health Program, 1994.

Appendix F

Reaching Spanish-speaking Workers and Employers with Occupational Safety and Health Information

Tom O'Connor, Consultant

INTRODUCTION

This paper addresses the occupational safety and health information needs of Spanish-speaking workers, their employers, and occupational safety and health practitioners , with a particular focus on the communication channels for reaching these audiences. Because other papers in this series are examining the background of occupational injuries and illnesses, and the present availability of Spanish language Occupational Safety and Health materials, this paper will not address these issues.

SPANISH-SPEAKING EMPLOYEES: DEFINING THE POPULATION

The employee population that is the subject of this inquiry is "Spanish-speaking workers." This covers foreign-born residents who have immigrated from Latin America and, to a much smaller extent, from Spain, as well as U.S.-born individuals who speak Spanish as their primary language. While we are referring not to "Hispanics/Latinos" but to "Spanish-speaking" individuals, little data exists specifically on the latter group. These two groups differ in two ways; first, the former includes a substantial segment for whom Spanish is not the primary language; second, the latter includes people whose country of origin is Spain. For this reason this paper will refer to data on the "Hispanic/Latino" population with the understanding that the two are not entirely equivalent.

In addition, while "Spanish-speaking" does not necessarily mean "immigrant," this paper will focus much attention on the Latino immigrant population for two reasons: (1) a high percentage of the population in this country whose primary language is Spanish are recent immigrants from Latin America; and (2) the recent immigrant population is at especially high risk for workplace injury and illness and, therefore, warrants particular attention in the process of developing strategies for Spanish-language Occupational Safety and Health training, education and outreach.

Number of Spanish-Speakers With Limited English

The 2000 census included a population survey that asked about individuals' use of Spanish as the primary language in the home. The survey resulted in an estimate of 18,520,000 Spanish speakers within the working age population range (18-64 years), with which we are primarily concerned in this paper.[1]

---

[1] U.S. Department of Commerce, Bureau of the Census, Census 2000 Supplementary Survey, Detailed Table 35: Age by Language Spoken at Home by Ability to Speak English for the Population 5 Years and Over. Supplementary Survey website: http://factfinder.census.gov/home/en/c2ss.html. Accessed June 19, 2002.

For the purposes of developing a strategy for the preparation and dissemination of Spanish-language materials, it is particularly important to know something about the English language abilities of this population. In other words, what is the population of Spanish-speaking workers whose English is insufficient to enable us to reach them effectively with Occupational Safety and Health materials in English? The census survey addressed this question, asking about the individual's ability to speak English and found the following results:

TABLE 1 Ability to Speak English

Population 18-64 Years Old	Estimate	Lower Bound*	Upper Bound*
Speak Spanish	18,519,675	18,339,602	18,699,748
Speak English "very well"	9,198,012	9,062,738	9,333,286
Speak English "well"	3,364,412	3,291,434	3,437,390
Speak English "not well"	3,787,984	3,694,688	3,881,280
Speak English "not at all"	2,169,267	2,082,805	2,255,729

*Using 90 percent confidence interval.

Thus, there are an estimated 18.5 million working age people in the United States. for whom Spanish is the primary language. It is perhaps surprising to find that some two-thirds of these individuals report being able to speak English "well" or "very well." The remaining approximately 5.96 million report speaking English "not well" or "not at all." It is also important to remember that there is a substantial population of immigrants working in the United States without legal documents, a high percentage of whom speak little or no English, and who were missed in the census count. Despite a more successful effort at a comprehensive count in 2000 than in previous censuses, a certain degree of under-counting among recent immigrants was inevitable.

An analysis of the records of 1,668 Hispanic immigrant job seekers at a community center in Durham, North Carolina, for example, found that a much lower proportion spoke good or fluent English than that found in the census. The analysis broke down as follows:

TABLE 2 Hispanic Community Job Information Center Clients, Durham, North Carolina

English Ability	Number	Percentage
None	609	36.5
Little	719	43.1
Good	129	7.8
Fluent	83	5.0

SOURCE: Data from the Hispanic community Job Information Center, at the Centro Hispano in Durham, North Carolina. Surveys conducted among job seekers from approximately January 2001 through March 2002. English ability was self-reported. Data from 20 African immigrants was excluded from these results.

It should be noted that this group of immigrants is primarily composed of recent immigrants from Mexico and Central America and may not be representative of the Latino population of the area as a whole. However, it *is* representative of the recent Latino immigrant population, which is likely to be working in the highest risk jobs in this country. The discrepancy regarding English-language capacity with the census figures—with less than 13 percent of this group speaking English well—is striking. We should consider this roughly 6-million figure from the census to be a significant underestimate of the actual number of Spanish speakers in the United States who speak little or no English.

DEMOGRAPHIC PROFILE OF THE SPANISH-SPEAKING WORKING AGE POPULATION IN THE UNITED STATES

Educational Level and Occupation

In order to consider the information needs of Spanish-speaking workers in the United States, we must look at the demographics of this population. The rapidly growing Spanish-speaking workforce in the United States is by no means monolithic and varies significantly by region. In the states that have been the traditional immigrant destinations (e.g., California, Texas, Florida, Illinois, New York, and New Jersey), for example, there is a large population of second-and third-generation immigrants for whom Spanish remains the primary language, who have attained a relatively high level of education and are employed in professional positions and skilled trades. Conversely, in the states that have only recently experienced major immigration flows, such as Arkansas, Georgia, and North Carolina, the great majority of Spanish speakers fit a similar demographic profile: They arrive with little formal education, speak little or no English, and are employed in jobs on bottom rung of the economic ladder. This latter group of states has experienced a tremendous growth in Latino population, with rates of increase from 1990 to 2000 between 200 percent and 400 percent for Georgia, North Carolina, Arkansas, and Tennessee, for example.

Even in these "new immigration" states, however, there is some diversity within the Spanish-speaking population. Deteriorating economic and political conditions in Central and South America have led to a significant increase in immigration among educated professionals, with varying levels of English-language literacy. It is not uncommon in many parts of the United States to find Colombian engineers working as warehouse stockers, Honduran doctors working as hospital orderlies, or Peruvian teachers assembling computers, for example.

The 2000 census reported that 43 percent of the Hispanic population of the United States had an educational level of less than a high school diploma, compared to 11 percent of the non-Hispanic population. Only 10.6 percent had a bachelor's degree or greater, compared to 28.1 percent of the non-Hispanic population. It should be noted, however, that these refer to individuals "of Hispanic origin," which is not equivalent to individuals whose primary language is Spanish and have limited English ability. If we narrow our focus to look at foreign-born U.S. residents from Latin America, we find that 34.6 percent have less than a ninth-grade education, and 50.4 percent have less than a high school diploma. A small study of Spanish-speaking construction workers by the North Carolina Occupational Safety and Health project in 2000 found that the median educational level was 7.5 years. Only 24 percent had completed high school.[2]

Census data on occupation indicate that Hispanics are over-represented in the categories of "operators, fabricators, and laborers" (22 percent of Hispanics compared to 13.4 percent of the general population) and service occupations (19.4 percent compared to 13.9 percent of the general population.) Again, these differences would be starker if we narrowed the focus to those whose primary language is Spanish.

In addition, an important factor affecting the health and safety of immigrant workers is the dislocation in occupational circumstances caused by immigration itself, as explained by Eduardo Siqueira, who has studied these issues extensively:

---

[2] *Immigrant Workers at Risk: A Qualitative Study of Hazards faced by Latino Immigrant Construction Workers in the Triangle Area of North Carolina,* North Carolina Occupational Safety and Health Project, June, 2000. *(Unpublished report)*

> "It is important to emphasize that immigrants often work in very different workplaces and situations when they migrate to the US. They may be peasants at home and become service sector or construction workers (which is worse) here. They may be young and inexperienced at home and are asked or end up working in dangerous jobs here without any training, safety or otherwise, whatsoever. They may also be middle class in their home country and end up working in blue-collar jobs here. Therefore, the real issue is how much work experience in the given industry they have before getting a job in it. There is often a large and sometimes drastic change in work environment conditions from their previous jobs, no matter what safety knowledge they had before. Anybody would be affected by this, not only immigrants."[3]

Age and Gender

The Hispanic population of the United States is substantially younger on average than the non-Hispanic population, with a median age of 25.8 years, a full 13 years less than the non-Hispanic white population. This is not only a function of more children in the average Hispanic family but also reflects a substantially higher proportion in the 18-24 year range, a group likely to be working and particularly vulnerable to workplace hazards.[4] Census data corroborate the widespread impression that a great many young men in their teens and twenties often with little prior work experience, come to the UnitedStates from Mexico and Central America seeking work. In addition, there is much anecdotal evidence that teenaged Hispanic males frequently add a few years to their reported age when applying for work, in order to gain jobs from which minors are barred. A recent study of young Latinos in construction found several teens who reported starting construction work in the U.S. at age 14 or 15.[5]

While Hispanic men outnumber Hispanic women in the United States, it is a smaller gap than is often believed. Females make up 48.5 percent of the Hispanic population of the United States and with a median age of 26.3 a great number of them are in the workforce. A recent report emphasizes that

> "available research appears to show that Hispanic women face greater risk of occupational injury and illness than non-Hispanic white women. This is due in large part to the disproportionate representation of Hispanic women in high-hazard industries and occupations. A 1989 California study showed Hispanic women experienced incidence rates of occupational injury and illness that were 1.5 times that of non-Hispanic white women. Because many Hispanic women are employed in "informal" industries or in industries where safety, health, and wage laws might not be routinely followed (e.g., "sweatshops" in the apparel, restaurant, food processing plants, or other industries), the risks can become even greater."[6]

---

[3] *Eduardo Siqueira, University of Massachusetts-Lowell, Work-Environment Program, personal communication.*

[4] U.S. Department of Commerce, Bureau of the Census, *Census 2000* Population by Age, Sex, Race, and Hispanic or Latino Origin for the United States. Supplementary Survey website: http://factfinder.census.gov/home/en/c2ss.html. Accessed June 19, 2002.

[5] North Carolina Occupational Safety and Health Project, unpublished data, 2002.

[6] Scott Richardson, *Hispanic Women and Occupational Health.* Paper presented at the first Hispanic forum on a safe and healthy environment October 18-19, 2000.

RECOMMENDATIONS

So what does this mean for the development of appropriate Spanish-language Occupational Safety and Health materials? It means that strategies for developing and disseminating such materials should take into account the following factors:

- The Spanish-speaking working population is diverse in such demographic characteristics as education, income, and occupation. Thus is there is some need for materials targeting individuals at a variety of literacy levels and in a variety of occupations.
- There is a substantial segment of the Spanish-speaking workforce that has little English language ability and, is most in need of Spanish-language Occupational Safety and Health education and training. This group is composed predominantly of recent immigrants with little formal education, occupying low-wage and often hazardous jobs in such industries as meat processing, construction, and textile and apparel manufacturing.
- The Spanish-language literacy level of this population is likely to be relatively low, on average, given the relatively low level of average education. This suggests that a high premium must be placed on developing low-literacy materials when designing Spanish-language materials and that approaches other than, or in addition to, written materials must be an essential element of a strategy to reach Spanish-speaking workers.
- Contrary to popular perceptions, Spanish-speaking women not only play an important part in the workforce, but also are exposed to significant workplace hazards. Efforts to reach Spanish-speaking workers to prevent occupational injuries and illnesses must include efforts to reach women workers.
- A high proportion of Hispanic workers are young and relatively new to the workforce. All the lessons learned in educating and training young workers in the United States must also be applied to the Spanish-speaking worker population.
- A substantial segment of the Spanish-speaking workforce is made up of immigrants who are working in hazardous jobs, such as construction, but who have little or no home-country experience in these fields, making their need for training even greater.

CHANNELS OF COMMUNICATION FOR REACHING WORKERS

In the Community

In the next few sections we will look at how information can best reach Spanish-speaking workers in the United States and some of the challenges that different approaches present.

In discussions of effective outreach programs to Hispanics in the United States a frequently recurring theme is the importance of trust—that the only way to reach Hispanics effectively is by first establishing a relationship of trust. It is often noted that representatives of government agencies have a particularly challenging task in reaching out to Hispanic immigrants in that these communities are likely to have a high level of suspicion of government representatives. This suggests that rather than government attempting to reach workers directly, it would be more effective to reach them through intermediary agencies that they trust. Some examples would be local Hispanic community centers, churches, immigrant advocacy organizations, non-profit worker advocates such as the COSH groups (committees on occupational safety and health), and unions. Some examples of effective worker outreach on Occupational Safety and Health issues through these types of channels include:

1. In Atlanta, an area OSHA office has partnered with the Roswell Intercultural Center, a Hispanic community organization that provides a wide range of services, including a major project serving day laborers in the construction industry. Day laborers are at high risk for occupational injuries because of their limited access to Occupational Safety and Health training and frequent mobility, among other reasons. The OSHA office has established a partnership in which OSHA staff, professional trainers from construction companies and fall-protection-equipment manufacturers, and local worker safety advocates come to the Roswell site and present full-day safety trainings. Workers are required to attend the training in order to gain access to the full services provided by the Roswell center.
2. In Durham, North Carolina, a COSH group teamed up with the local Centro Hispano to create a Job Information Center in which recent immigrants would gain access to information about jobs. A condition of receiving the job information is that all participants attend an orientation training that includes basic job safety and health principles and information on workers' rights under OSHA law. For the COSH group the goal was to provide Occupational Safety and Health training and education, but this could be accomplished effectively only by providing the community with what they most wanted and needed: access to job information. Training is conducted in Spanish by native speakers (sometimes in conjunction with a co-teacher who is not a native speaker), with individual English translation provided for the occasional non-Spanish speaking participant.
3. The Santa Clara (California) Committee on Occupational Safety and Health has developed a successful occupational safety and health training program for immigrant women working in the electronics industry in Silicon Valley. The program, called "WE-LEAP" (the Working Women's Leadership project) brings together groups of women of various language and ethnic groupings, and provides them both practical training on a variety of job-related topics of interest to them and Occupational Safety and Health training. Training sessions use a variety of modes that are grounded in the culture of the group being trained: story telling, drawing and painting, rituals, dances, and songs. The training uses a four-part "worker story process" that takes a holistic approach to the issue of occupational safety and health that includes gender and family issues in the discussion. The workshops end with a presentation by the trainees of what they learned, again using their own cultural modes of expression. The training sessions provide a social outlet for the women as well as important information about health and safety and workers' rights on the job.
4. The San Francisco Department of Health developed a successful partnership with a number of local agencies to conduct an outreach and educational program for day laborers, most of whom are Spanish-speaking. Health department staff and other professional trainers conduct training at locations where the day laborers regularly congregate, such as a local day-labor center, homeless shelters, a church, and on the street corners themselves. This project did involve a government agency, thereby overcoming the obstacle of immigrants' fears of government. By establishing a regular presence in the community, the government agency staff were able to gain the trust of the community and gain good access to a highly vulnerable population.
5. Around the country many unions have recognized that providing services desired by immigrants is the most effective way to bring these workers to them and give them greater access to health and safety and other training. The UNITE textile workers union local in New York City, for example, offers a wide range of services, such as English classes and immigration assistance, to its multi-lingual immigrant membership.

These models illustrate a few key principles that are applicable to all worker training but are particularly useful with non-English-speaking workers.

- *Provide workers what they want and you may get an opportunity to reach them with the information that you want to provide.* It is difficult to attract workers in general to programs that are primarily focused on occupational safety and health. It is simply not a high priority for most workers to increase their knowledge of these issues. This is particularly true among Spanish-speaking workers in the United States, who are likely to work longer hours, have less free time to attend training programs, and are likely to be more mistrusting of anyone perceived as an outsider.
- *Go out and find workers where they are.* We can't simply produce training materials, file them in our libraries, and expect workers (or employers) to seek them out. An officer of a state Department of Labor wage and hour section once told me that the section didn't believe there was a problem with non-payment of wages to Hispanic workers because they didn't receive many calls from Spanish-speaking workers. I pointed out that my non-profit agency received hundreds of calls alleging non-payment of wages, because we had made a point of getting information on this issue out into the community. Only by actively going out into communities and reaching out to workers where they are can we effectively reach Spanish-speaking workers. Some outreach programs have effectively distributed information at soccer games, community festivals, Hispanic neighborhoods and trailer parks, and other venues where the Hispanic community gathers.
- *Make training sessions interesting.* Providing an enjoyable, comfortable environment for training creates much greater opportunities for learning and will bring people back in the future.
- *Use Hispanic trainers who are native speakers, when possible.* While non-Hispanics may be perfectly capable of providing Spanish-language training, there is no avoiding the fact that immigrants, in particular, are most comfortable receiving information and participating in training in which the trainer is of their ethnic background. It is sometimes possible to use a team-teaching approach with a Hispanic trainer paired with a non-Hispanic.
- *Avoid if at all possible providing training in English that is then translated into Spanish.* This process can be used as a last resort but is far inferior as a training method. For one reason, only half as much material can be covered because of the time taken in interpreting. Second, some content is invariably lost in translation and the instructor will be unable to facilitate any discussion among the group. This type of training inevitably is restricted to a simple lecture format that discourages much participation.
- *Provide childcare to facilitate the participation of women workers.* In the vast majority of Hispanic families childcare is taken care of within the family. If another family caregiver is not available, it is common for Hispanic women to attend events accompanied by their children. Providing child care at training sites not only increases the likelihood that women workers will be able to attend but also increases their ability to participate in the sessions without the distractions of their children's presence.
- *Be prepared to use low-tech training methods.* Most community training venues are distinctly low-tech. Trainers in these venues should not expect to do Power Point presentations or have other advanced audiovisual equipment available.

Through Unions

There is a widespread perception that Spanish-speaking workers, particularly recent immigrants, are less likely than other U.S. workers to join unions and to turn to them for information and assistance. While this may be true for the most recent immigrants, many of whom lack legal work documents and are thus reluctant to draw attention to themselves, it is not necessarily true for Hispanic workers as a whole. It is commonly noted that recent immigrants from Mexico have negative attitudes about unions because of the long history of corruption and abuse of workers by "official" Mexican unions tied to the political establishment. What is often overlooked is the presence in this country of a substantial population of immigrants from Central and South America, where unions have often played central roles in promoting the welfare of working families and fighting, literally and figuratively, for workers' rights against repressive military-backed regimes. These immigrants, according to many observers, are more likely than the average U.S. worker to trust unions and to want to join them. One survey found that in the 1998 California referendum on the anti-union Proposition 226, 75 percent of Latinos opposed the measure, compared to only 53.5 percent of non-Latinos.[7]

Many unions have, in fact, established effective health and safety training programs for Latino immigrants and have produced quality training and educational materials in Spanish. Some of the international unions that have been particularly active in this area are the United Food and Commercial Workers (UFCW), UNITE textile and garment workers union, United Auto Workers (UAW), Laborers' International Union, and the Retail Wholesale and Department Store Workers Union (RWDSU).

The Internet as a Channel of Communication

More and more, business, government, and non-profit agencies are using the Internet to disseminate information. The Internet provides a tremendous resource for reaching countless people with our messages and offers a wonderfully flexible forum for doing so. Just 10 years ago it was difficult for a non-specialist to know how to find chemical hazard information, for example; today a few keystrokes on an Internet search engine provide a mind-boggling wealth of information. For those who have good access to the Internet this is a rich treasure of resources.

We must not forget, though, in our excitement over the possibilities of the Internet that many Americans still lack reliable access to this resource. This is particularly true of low-income workers and Spanish speakers. A recent Commerce Department report found that while Hispanic Americans' use of the Internet is growing fast—a 30 percent growth rate from 2000 to 2001—it still lags far behind other groups. Only 31.6 percent of Hispanics use the Internet, compared to about 60 percent of Asian Americans and 60 percent of whites.[8] An outreach and education strategy that primarily relies on the Internet will leave out nearly 70 percent of the Hispanic community at this time. If we consider that those recent immigrants who speak little or no English are even less likely to use the Internet, we can assume that an Internet-focused strategy would miss the vast majority of the highest-risk Spanish-speaking workers.

---

[7] Bailey, Eric, and Shogan, Robert. Defeat of Measure Energizes Labor. Los Angeles Times, June 4, A3, A28. Cited in Organizing Immigrants: The Challenge for Unions in Contemporary California, edited by Ruth Milkman, p. 8. (Ithaca, NY, Cornell University Press).

[8] A Nation Online: How Americans are expanding their use of the Internet, U.S. Department of Commerce, Feb. 2002. (National Telecommunications and Information Administration, U.S. Department of Commerce, Washington, D.C., 2002).

The Worksite: Limitations of Worksite-based Training

Training at the worksite has some distinct advantages. Workers are a "captive audience," that is, they can be required to attend training that is on company time, guaranteeing a good turnout. In addition, training employees of a single employer ensures that the training will be focused appropriately and specifically to the needs of that workforce. The employer's buy-in to the training, allowing it to take place at the worksite, increases the likelihood that it will lead to substantive changes to prevent hazardous conditions in the workplace.

Worksite training, however, has a number of major drawbacks, which can become even more pronounced when conducting training for low-wage immigrant workers. These include:

- *Open dialogue is stifled by the presence of management.* Many if not most workers of all backgrounds in this country are reluctant to speak up about workplace concerns, such as health and safety hazards, in the presence of company managers. This is particularly true of immigrant workers who are especially fearful of the consequences of losing their jobs and whose culture is often deferential to those in superior positions in the workplace. In these settings it is difficult to elicit any critical commentary or concerns from workers about safety and health conditions in the workplace.
- *Competition with work demands.* When training is not removed from the worksite, it is often disrupted by the demands of the work. Worksite trainers may arrive for their scheduled training on time at 9:00 a.m. only to find that production has been stepped up and the line employees can't be spared until 11:00 a.m.and the training, therefore, will only be an hour instead of the promised three hours. Breakdowns of machinery or other problems on site often require key participants to leave in the midst of training.
- *Logistical Obstacles.* Training conditions are frequently less than ideal. Training may take place in the employee lunchroom, for example, with the trainer forced to make herself heard over the din of a crowd of non-participating employees eating lunch. Acoustics may be difficult. Managers may crowd a full shift of 100 employees into a room in order to get everyone into a training as quickly and efficiently as possible, making effective interaction impossible.

Spanish Language Media

While the number of English-language daily newspapers in the United States has been steadily declining in recent years, the number of Spanish-language dailies in this country grew from 14 in 1990 to 34 in 2000 and the number of weeklies increased from 152 to 265. Magazines doubled from 177 to 352.[9]

The Allied Media Hispanic Publications Network reports that 91percent of U.S. Hispanics speak Spanish at home and 67percent are more comfortable with Spanish-language publications, suggesting that Spanish-language newspapers could potentially be an effective method of reaching this population.[10]

As noted above, the educational and literacy levels of a substantial segment of the Latino immigrant worker population in the United States are limited. Thus, it is critical to reach out to these groups using strategies that do not rely exclusively on written materials. Spanish-language television

---

[9] Data from the National Hispanic Media Directory cited in K. Campbell, Demographics drive the Latino media story. *The Christian Science Monitor*, June 21, 2001 (p 14).

[10] CASS Hispanic Publication Network. Information <casscom.com/ethnic/hispanic.html>. Accessed June 19, 2002.

and radio offer excellent opportunities to achieve this. Several Spanish-language TV networks broadcast nationally, including Telemundo, Univision, Azteca America, and Telefutura, and have huge viewing audiences among the Spanish-speaking population. In addition, there are currently 594 U.S. radio stations that broadcast in Spanish. These radio and television outlets represent other powerful potential outreach channels for occupational safety and health information targeting Spanish speakers.

WHAT ARE THE PRIORITY CONTENT NEEDS FOR SPANISH-LANGUAGE WORKER TRAINING AND EDUCATIONAL MATERIALS?

Needs for Spanish-language Occupational Safety and Health materials and training can be divided into four general areas.

1. Basic information on OSHA and workers' rights under OSHA standards
2. Basic "hazard awareness"
3. Skills for addressing work hazards and protecting one's health and safety on the job
4. Industry- or hazard-specific materials and training

Basic Information on OSHA and Workers' Rights

A 2000 study of Latino construction workers in North Carolina found that the workers had a low level of knowledge of basic health and safety laws in this country. Participants in the study were asked, "If you thought that there was a dangerous situation at work and the boss wasn't doing anything to correct the problem and you wanted to make a complaint about the situation, do you know what you could do to make a complaint?" Only 1 of the 45 respondents named the Department of Labor, while not a single respondent mentioned OSHA. After responding to this question participants were asked if they had ever heard of OSHA. Only 15 out of 43 (35 percent) said they had heard of OSHA and were able to explain something about its role.[11]

This study supports a contention frequently made by educators and advocates who work with recent immigrants that the greatest need among this population is basic education on OSHA law and workers' rights to safe and healthy conditions under these laws. A number of Occupational Safety and Health educators who work with Spanish-speaking workers noted that the greatest need for training among the immigrant worker population is not technical information but training on how to use the rights that OSHA gives them to protect their health and safety on the job. One educator who works with meat processing workers commented:

> "Training to shift power - it's important to include that workers are not vulnerable because of race or immigrant status. Workers in general wield less power in the workplace and this power differential is especially acute among immigrant and minority workers for a number of reasons. However training should not only be about increasing knowledge among workers, but about changing this power differential."[12]

---

[11] *Immigrant Workers at Risk: A Qualitative Study of Hazards Faced by Latino Immigrant Construction Workers in the Triangle Area of North Carolina*, North Carolina Occupational Safety and Health Project, June, 2000 (unpublished report).

[12] Susan Cameron, United Food and Commercial Workers Union, personal communication.

Hazard Awareness

Clearly, training on workers' rights under OSHA, however, is not sufficient in itself. Spanish-speaking workers with little prior safety and health training often are in need of basic health and safety "hazard awareness" training,to raise their level of awareness of the existence of hazards in their workplaces that they may not recognize as the potential source of serious health problems (e.g., chemical and ergonomic hazards) or that they may simply see as unavoidably "part of the job." A number of good tools and materials are currently available in Spanish for this type of hazard awareness training from sources such as the UC, Berkeley's, Labor Occupational Health program, the UCLA Labor Occupational Safety and Health program, COSH groups, and unions. The reader is advised to consult Marianne Brown's white paper in this series, which addresses this in greater detail.

Skills in Addressing Work Hazards

In addition to knowledge about work hazards there is also a great need for training of Latino immigrant workers on how to address those hazards. A number of useful training materials and techniques have been developed for this purpose, some of which are currently available in Spanish; others would need to be translated.

It is important that this training focus on the realities that immigrant workers face. We must recognize that many of these workers are fearful of making complaints for a variety of reasons. Simply providing them with information about the hazards and about their rights under OSHA to protect themselves from these hazards misses the key point that this information may not result in any changes. Training should include a focus on real-life problem solving in the workplace, given the reality that many workers will not be willing to make a complaint and that many of them believe, correctly, that the OSHA laws that are intended to protect them from retaliation often fail to do so. A number of simple small group exercises have been developed that pose problem situations and ask participants to develop solutions. This type of exercise is valuable in encouraging immigrant workers to consider their options when faced with a hazardous work situation and to work on developing creative solutions.

Industry- and Hazard-Specific Materials and Training

Finally, there is a need for materials and training specifically targeting industries and hazards in which Spanish-speaking workers predominate. The priorities in terms of the content of this training should be based upon the results of the analysis of injury and illness data among Hispanic workers that is being reported in another white paper in this series. We can assume, however, that priority target industries will include construction, meat processing, garment and textile industries, and agriculture, to name a few of the most likely targets. As noted earlier in this paper, it is important to keep in mind the varied audiences for these materials, ranging from educated professionals to very-low literacy workers.

LITERACY ISSUES

In designing materials for Spanish-speaking workers it is not enough to simply take existing materials and translate them into Spanish. This may serve the purposes of a certain segment of the population—the more educated segment—but it will fail to meet the needs of those workers who most likely will be in high-hazard jobs. While no hard data is available on Spanish-language literacy among Spanish speakers in the United States, we do know that 13 percent of the population of Mexico, the largest source of Spanish-speaking immigrants to the United States is illiterate in Spanish.[13] Given the fact that poor, less educated, rural residents of Mexico are the most likely to emigrate to the United States, the proportion of Mexican immigrants who are illiterate in Spanish is probably higher than this.

It bears mentioning that this lesson regarding literacy is an important one for English speakers as well as Spanish speakers. The 1992 National Adult Literacy Survey (NALS) found that between 21 percent and 23 percent of U.S. adults are functioning at the lowest level of literacy. At most, people at this level are able to perform tasks involving a brief, uncomplicated text, but many do so with difficulty. An additional 25 percent to 28 percent of the participants are functioning at Level 2, which the Department of Education describes as "more varied than those at Level 1 but still quite limited." They have "considerable difficulty carrying out tasks requiring them to use long texts or do 2-step calculations."[14]

Most Occupational Safety and Health documents currently available for workers, in English and in Spanish, are written at too high a reading level to enable the majority of workers to comprehend the information. One study found that the average worker cannot understand 40 percent of the content of the information on Material Safety Data Sheets (MSDS).[15] Another analysis done for OSHA found that the average MSDS was written at a college level, well above the reading level of most workers.[16] Similarly, the Labor Occupational Health program of UC, Berkeley, conducted a review of 25 health and safety materials produced by government agencies, unions, educators, and companies and found that the average reading level was college level. Only four of the samples were at or below the eight-grade level, the level that the program recommends for widespread comprehension.[17]

This trend is true not only for printed material but also for website content. One study that examined the suitability of website content for low-literacy, non-English-speaking users concluded that "perhaps the greatest gap we found in content is material for the 44 million adults in the United States who lack functional literacy skills to perform everyday tasks. Of the 1,000 sites we reviewed we found only 10 that were appropriate for limited-literacy adults."[18]

There are a number of good resources available to assist Occupational Safety and Health trainers in preparing low-literacy materials. These include *Teaching About Job Hazards: A Guide for Workers and their Health Providers* by Nina Wallerstein and Harriet Rubenstein. This book provides an excellent overview of effective adult education and training methods, guidelines for providing education during screening programs, preparing factsheets and training materials, and evaluating health and safety education. The UC, Berkeley, Labor Occupational Health program's excellent book *The Right to Understand: Linking Literacy to Health and Safety Training* also provides an extensive guide to effective training and materials development for low-literacy workers. The latter book provides a good summary of the techniques for effective materials development for low-literacy audiences. These principles should be kept in mind when preparing Spanish-language materials for workers in the U.S.

---

[13] CIA World Factbook 2002. *www.cia.gov/cia/publications/factbook/geos/mx.html*. Accessed June 19, 2002.

[14] Lynn Jenkins and Stephane Baldi, *1992. Adult Literacy in America, U.S. Department of Education.* National Adult Literacy Survey (US Department of Education, Washington, D.C.).

[15] *"Rights and Realities: A Critical Review of the Accessibility of Information on Hazardous Chemicals." Sattler, Barbara. Occupational Medicine: State of the Art Reviews, April-June 1992 (*pp. 189-196).

[16] *The Right to Understand: Linking Literacy to Health and Safety Training.* By Elizabeth Szudy and Michele Gonalez Arroyo, Labor Occupational Health Program, UC-Berkeley (1994, Berkeley, CA).

[17] *Ibid*, p. 33.

[18] "Online Content for Low-Income and Underserved Americans." The Children's Partnership. Undated. <http://www.childrenspartnership.org/pub/low_income/>. Accessed June 19, 2002.

Writing

- Establish your priority message.
- Organize text into short, logical sections.
- Use words that are easy to understand.
- Define technical terms.
- Keep sentences short and simple.
- Use a conversational style and active voice.

Design

- Use large type.
- Emphasize important points by underlining, bold type, italics, and boxes.
- Use wide margins.

Illustrations

- Use simple line drawings.
- Illustrate the correct way to do things, not the wrong way.
- Avoid abstract graphs and charts.

(Adapted from *The Right to Understand: Linking Literacy to Health and Safety Training*. By Elizabeth Szudy and Michele Gonalez Arroyo, Labor Occupational Health Program, UC, Berkeley.)

Rather than repeating all the lessons from the resources cited above the reader is encouraged to consult them for further guidance on producing effective materials for low-literacy learners.

TRANSLATION ISSUES

In the experience of this author and that of a number of other experts consulted for this paper, the quality of available Spanish-language Occupational Safety and Health materials is mixed. A review of Spanish-language Occupational Safety and Health documents currently available on the Internet found that the majority are of good quality, but there are some notable exceptions. In some cases translations clearly were not checked by a native speaker and the results are confusing and even misleading. For example, a translation of a fact sheet on the OSHA fall protection standard states in the Spanish version that the standard has been effective, as in "successful," since a given date; the original meaning was that the standard has been "in effect" since that date. Equally confusing and misleading errors occur throughout the document. Similarly, a Spanish-language guidance document recently placed on the OSHA website was filled with errors and garbled language. (The NIOSH Spanish-language website is of good quality.)

One expert consulted on this issue contends that most available Spanish-language Occupational Safety and Health training materials in the United States are of inferior quality. Dr. Fernando Marroquin of the University of Alabama suggests that the problem is a combination of two factors: Many of the people doing these translations are not actually fully literate in Spanish and the translators are often not familiar with Occupational Safety and Health terminology. He suggests that the first problem is caused by translations that are often done by second-generation Latinos in this country whose Spanish is learned haphazardly "on the street" and is inadequate to do complex technical translations.

Country-Specific Spanish: Not a Major Barrier

Non-specialists often refer to the differences in Spanish language usage in different countries, suggesting that it is nearly impossible to create materials that are readable by all Spanish speakers. While regional and national differences certainly exist, these differences are small in relation to the commonalities. If materials are produced using "standard" Spanish (i.e., avoiding local idiomatic expressions or words used only in specific countries), it is not difficult to create materials in a language understood by all Spanish speakers. As one expert put it, "An educated writer from any Spanish-speaking country is perfectly capable of expression, on any subject, that is clear and unambiguous to an educated reader from any other Spanish-speaking country. Period."[19]

The Use of "Spanglish"

The substantial presence in the United States of Mexican-born immigrants has resulted in the development of an entire vocabulary of hybrid words mixing Spanish and English. Many of these are work-related terms, which immigrant workers may know only in "Spanglish" or English but not in Spanish. For example, in construction work Latino immigrants, particularly Mexicans, will often refer to the *finisheros* and the *chitroqueros* to describe the finish carpenters and sheetrock installers.

Some translators insist that these neologisms are not legitimate Spanish words and should never be used in educational materials. Others, myself included, believe that the most important thing in developing educational materials is to ensure comprehension. If you need to use a word that is not accepted by the Real Academia in Madrid but is the word used by your entire target audience to describe a given concept, you are better off using that word. Because many Latino immigrant workers learn a trade in the United States, they may know many of the English terms for work-related concepts but not the Spanish words. Thus, it may be useful in some cases to use both the Spanish and English words for equipment, job functions, etc. It does little good, for example, to refer in a document to madera contrachapada if every Latino construction worker in the U.S. knows it as *el plywood.*

Machine Translation: Not a Reliable Option at This Time.

High hopes have been placed on the possibility that computer-based machine translation systems could take on much of the burden of written translations. These expectations, however, have so far failed to materialize for the most part. A number of software programs for this purpose are widely available, but the results are uniformly unsatisfactory. As the programs themselves warn users, they are useful only in to giving a general idea of the meaning, not to provide a precise translation. The translated output can, in fact, provide an idea of the content, but it can be painful to read and can be misleading. The following is Spanish to English translation of a sample of text from of a NIOSH document:

> *In addition to the injuries, the materials and dangerous conditions of work also constitute a preoccupation for the adolescent workers. It is known less on this field than on the effects of the injuries (that have an immediate impact and that are possible to be counted and to be classified as far as the cause). The dangerous exhibitions of adolescent workers to materials and conditions of work could be in an immediate disease; nevertheless, it is possible that the disease cannot be detected per*

---

[19] *"Alli no se habla español" on the website of Contact International-the Center for Technical Translation at <http://www.cicenter.com/a_espanol.htm>*. Accessed on June 19, 2002.

> *months or years after the exhibition. The adolescent workers could be exposed to pesticidas in the work of farm and taking care of the turf, benzene in powerboats, lead in the adjustment of bodies, asbestos and silica in the work of construction and maintenance, and high levels of noise in the manufacturing industry, the construction, and also preoccupation by the possibility that has arisen the fatigue to study and to work could contribute to injuries between adolescent workers.*

The use of translation software to translate Web pages has also been much touted on the Internet recently. This application appears to be even less useful, at this point, than the use of software to translate blocks of text. The translated output is often confusing to the point of being unreadable. A typical example, a Spanish to English translation of a NIOSH Web page, follows:

> *The reactions begin of ordinary to the few minutes of the exhibition to latex, but hours can happen later and can produce different symptoms. The slight reactions present/display reddening, irritation, or picazón to the skin. Acute reactions can include respiratory symptoms such as nasal secretion, estornudos, picazón to the eyes or throat and asthma (difficulty to breathe, periods of cough and jadeo). In rare occasions, a shock state can take place; but a reaction that puts in danger the rare life time is the first symptom of the allergy to látex. These reactions are similar to the observed ones in some allergic people after undergoing a bee puncture.*

Clearly, the software at this point in its development is not useful as a substitute for human translation. Some proponents of machine translation have argued that these systems can be effectively used in conjunction with human translators, doing the "heavy lifting" of rough translations of long documents, which can then be polished by the human reviewer.[20] This may be of some value in settings like United Nations or European Union offices, where thousands of pages must be translated daily. It is debatable whether the output of the machine translators is, in fact, better than nothing or if it requires more time and effort to fix errors in sentence structure and translation than it saves time in providing correctly translated words.

LIMITATIONS OF WRITTEN INFORMATION ALONE

Most of the people who will read this paper and participate in this conference are, like myself, focused on the written word as the primary means of transmitting and gathering information. When we want to get a message across to a segment of the population, our first thought is to put it into writing. Until recently our next step would have been to have this information printed as fact sheets or reports. Now we are equally inclined to publish this information on websites, making it instantly available to millions of readers, but we tend to forget that the majority of the population do not, in fact, get their information from written sources. Newspaper readership has declined dramatically in recent years, while the number of Americans who get their news primarily from television is at record high levels.

Adult educators commonly point to the maxim that we all learn best when we receive a message through a number of channels—seeing and hearing it—and when we have the opportunity to think critically about the message and put it into practice in some form. Written materials by themselves, whether published as printed fact sheets or posted on websites, are unlikely to reach a broad audience. They are most effective when used as part of a broader communication strategy that

---

[20] *Cambridge Encyclopaedia of Language,* David Crystal, editor, p. 353. (Cambridge, England; Cambridge University Press, 1997)

includes group training in community, union, or workplace settings or in conjunction with one-on-one patient education.

Written materials may not be the most important priority for the Spanish-speaking workers at highest risk for occupational injury and illness. As one worker advocate commented in considering the question of what is most needed for Latino immigrant workers,

> "So what's the answer? Producing more materials is not the real solution (although it would help). There are lots of materials out there now. The issue is outreach. Immigrant and undocumented workers are not likely to find their way to the OSHA Web page or call up a U.S. government agency when they have a problem. They are comfortable in their community organizations, churches, etc., and with people from their community. This is why OSHA needs to provide funding to organizations that have the ability and knowledge to reach workers where they are comfortable."[21]

Jim Platner of the Center to Protect Workers' Rights, who has studied construction safety issues for many years, emphasizes that written materials alone are insufficient and that materials must be supplemented by on-the-job safety training.

> "Availability of translators or bilingual co-workers might be of little use when someone yells "look out below," as something falls off the catwalk over your head. While training materials are increasingly available in Spanish, critical skills are learned by observation of co-workers or journey-level workers who know the job, practicing the job with critical evaluation of performance, close supervision when you are new to a job, and other applied learning experiences. Translating these learning experiences must go beyond translating textbooks and fact sheets in order to successfully prevent occupational injury and disease in construction. While no specific research on the contribution of language issues to falls in Hispanic construction workers was found, it appears to be an important area for further attention and research. For example, the top cause of falls among Hispanic construction workers involves falls from scaffolds that are being put up or taken down. A closer look at this task might reveal that communication problems or training deficiencies could be contributing to the resulting injuries."[22]

SELECTING EFFECTIVE TRAINERS

In this author's experience and in that of a number of people consulted for this paper, a common story is heard about worksite-based training in Spanish. In many cases a safety professional or other company manager presents safety information in English to the employees, which is then interpreted in Spanish for the Spanish-speaking employees by a Latino employee who is bilingual. Or the bilingual employee may simply be given material in English and told to pass this information on to the Spanish speakers. This intermediary, in the vast majority of cases, has no safety and health background and no education or training experience.In organizations with large Spanish-speaking workforces these bilingual Latino intermediaries may be designated as the link between management and the Latino employees and take on important roles in the organizations, sometimes being assigned

---

[21] *Jordan Barab, AFL-CIO Health and Safety Department, personal communication.*

[22] Jim Platner "Language and communication problems might contribute to the risks faced by Hispanic Workers," Conference Report from the First Hispanic Forum on a Safe and Healthy Environment, available at <http://www.geocities.com/hispanic_eosh/doc.html>. Accessed on June 19, 2002.

tasks and given responsibilities far beyond their training. In these circumstances the English-speaking trainer has no assurance that the content workers are receiving is accurate or complete.

English-speaking trainers may also conduct oral presentations at the worksite in English and then pass out written information in Spanish, as in the case reported by this Latino construction worker:

> "They give us a 'safety' every week, every Monday. A paper comes around in Spanish and in English. The supervisors read it out loud in English and give us the Spanish one to read. And we sign it."[23]

While we should recognize that these employers are at least making some effort to provide safety information, which is more than what some employers do, this approach is likely to have limited effectiveness.

EMPLOYERS' NEEDS FOR SPANISH LANGUAGE INFORMATION

Defining the Population: Spanish-speaking Employers

A 1997 Census Bureau survey of business owners gives us a fairly detailed profile of Spanish-speaking business owners in the United States (including Spaniards). Four states, California (336,400), Texas (240,400), Florida (193,900) and New York (104,200), accounted for 73 percent of these firms.

Table 3 includes those industry sectors in which Hispanics employ a substantial number of people.

TABLE 3 Hispanic-owned Business with at Least One Employee by Sector.

Industry Sector	Firms	Employees
Construction	31,478	168,873
Manufacturing	10,173	171,738
Retail trade	48,713	324,474
Wholesale trade	14,125	94,281
Service industries	70,838	463,889

SOURCE: U.S. Department of Commerce, U.S. Census Bureau 1997 Survey of Minority-Owned Business Enterprises: Hispanic, Washington, D.C.

In addition, Hispanics own 5,925 businesses in the agricultural services, forestry, and fishing sector, employing 25,955 people. For our purposes in assessing needs for Spanish-language materials and training it is interesting to note the large number of employees in both the construction and the manufacturing sectors in Hispanic-owned businesses. Unfortunately, we have no data as to the English-language abilities of these business owners or that of their employees. We do know by ample anecdotal evidence from a wide range of sources around the country that Hispanic business owners in

---

[23] Quote from an interview reported in Immigrant Workers at Risk: A Qualitative Study of Hazards faced by Latino Immigrant Construction Workers in the Triangle Area of North Carolina, North Carolina Occupational Safety and Health Project, June, 2000. (Unpublished report)

the construction industry are most likely to hire a predominantly Hispanic workforce. These census data verify a widely held perception about these owners that they are concentrated in the "special trade contractors" sector. This sub-group accounts for approximately 80 percent of the construction firms owned by Hispanics (or 25,110 firms) and employs 120,791 people. Again, we do not have a count by language ability or ethnicity of this group of employees, but we can assume that at least a large proportion is Spanish-speaking.

In the manufacturing sector there is no particular pattern, with Hispanic ownership spread across a range of sub-categories. In the service sector two categories with significant workplace hazards are notably represented: auto repair and services, with 11,662 firms employing 43,534 people, and health services, with 16,900 firms employing 96,349 people. Table 4 indicates the breakdown of these business owners by country of origin.

TABLE 4 Hispanic-Owned Firms by Country of Origin, 1997

Ethnic Origin	Firms	Receipts (000s)
Total	1,199,900	186,300
Mexican	472,000	73,700
Cuban	125,300	26,500
Unspecified/other*	475,800	61,700
Puerto Rican	69,700	7,500
Spaniard	57,200	16,900

*This category includes unspecified written-in "Hispanic Latin-American" and "Other Hispanic" responses.

Recommendations for Reaching Spanish-Speaking Employers

What does this mean for planning strategies for developing and disseminating Occupational Safety and Health materials in Spanish aimed at business owners? This is somewhat difficult to interpret in that we don't know with any certainty the extent that Spanish-language materials are needed by these business owners (without knowing their English-language capacities.) Anecdotal evidence suggests that a high percentage of Hispanic employers in the construction industry may not be fluent in English and would benefit from Spanish-language materials. Thus, it would be advisable to include a focus on the construction industry when developing materials in Spanish for employers. The above data also suggests that the auto service and repair industry may be an area where many employers could benefit from Spanish-language materials.

A report from the *First Hispanic Forum on a Safe and Healthy Environment*, in fact, included as one of its principal recommendations to target and involve Hispanic construction contractors in Occupational Safety and Health. The report noted that "Hispanic workers in construction include managers. Although some may be self-employed and combine management with production work, managers were among the top five Hispanic construction occupations by number. Hispanic managers and contractors are an important group for future partnering. They are likely partners for Spanish-language safety and health materials, and they can help to develop, test, and disseminate best practices to help raise the standard of safety practice in the industry."[24]

The census data cited above support the contention that Hispanic contractors form a significant subset of the construction industry. A number of people consulted for this paper reiterated

[24] Conference Report-Actions, from the First Hispanic Forum on a Safe and Healthy Environment, available on the conference website at <http://www.geocities.com/hispanic_eosh/doc.html>. Accessed on June 19, 2002.

the above comments regarding the importance of reaching out to Hispanic construction managers. There is a widespread perception that Hispanic-owned construction companies are more likely to cut corners on safety. This perception was supported by interviews with Latino construction workers in North Carolina, a number of whom stated that in their experience, Latino employers were worse to work for than non-Latinos.[25]

The North Carolina study and discussions with a number of experts for this paper also both pointed to the key role played by a person commonly found in the construction industry: the bilingual Latino supervisor. These individuals often play a role as trainers (to the extent that on-the-job training is done), interpreters, and intermediaries between workers and management. They are in a good position to reach their Hispanic employees because of language and cultural commonalities. At the same time, however, they may have received little or no training in safety and health. A former OSHA compliance officer reports that she has inspected many incidents involving Latino worker fatalities and found that in many cases, a language barrier was not involved—that the supervisor was Latino. But, she found, the supervisor was often poorly trained and ill-equipped to carry out the functions of a supervisor, including overseeing safety and health conditions.[26]

Training of these bilingual supervisors, particularly in the construction industry, could have a significant impact. Training should emphasize the supervisor's responsibility for ensuring safe and healthy working conditions. In addition, training should aim to overcome the "macho" attitudes that are commonly found among construction workers, even more so among Latinos. An OSHA compliance assistant tells of her experience conducting a training on scaffolding safety in[27] which the workers told her that the male instructor was a *maricon*, a sissy, because he told them they should never work on a scaffold that was unstable. They told her that the other workers would laugh at them if they appeared to be afraid. This *machismo* presents an obstacle that must be addressed in training.

Many English-speaking supervisors and foremen receive this training at local community colleges, but few programs are available in Spanish. Providing more of these training programs in Spanish would meet an important need.

Train-the-Trainer programs

A number of studies and reports have indicated that Latino workers are particularly likely to learn about safety from their co-workers rather than from formal training programs. This suggests that there is great potential in training supervisors and other Latino workers to train other workers on health and safety issues. The advantages of these train-the-trainer programs are many: Workers are more likely to listen to and accept information from those they trust; training can be ongoing, rather than one-time; and training tends to be more grounded in the reality of the workplace. Although these programs require a larger additional investment, they can pay great dividends in preventing injuries and illnesses on the job in the long run.

---

[25] *Immigrant Workers at Risk: A Qualitative Study of Hazards faced by Latino Immigrant Construction Workers in the Triangle Area of North Carolina, North Carolina Occupational Safety and Health Project.* (Cambridge, England; Cambridge University Press, 1997).

[26] Marilyn Velez, former OSHA compliance officer, currently a compliance assistant with OSHA's Atlanta office, personal communication.

[27] Ibid from footnote 27.

Appendix G

An Examination of the Occupational Risks and Occupational Safety and Health Communication Needs of Spanish-Speaking Children Who Are Employed or Live on Farms

Martha Soledad Vela-Acosta, MD, MS, Ph.D.,
University of Texas, Health Sciences Center,
School of Public Health at Brownsville

Acknowledgment: My gratitude to Jimmy Perkins, CIH, Ph.D.

INTRODUCTION

Children's health relies on the interaction between their environment (living and working conditions) and their developmental status. While agriculture provides work opportunities for children worldwide, these opportunities also lead to potential risks. Many Spanish-speaking children are working as farm laborers, and face a range of common occupational agricultural risks. They form an emerging occupational group in the United States agricultural industry. Children are present as family members on farms operated by farmers of Hispanic[1] origin, as nomadic adolescent farmworkers[2] traveling solo, and as part of migrant farmworker groups either as family members or as individuals. While the first two groups are increasing in number, health risks and outcomes are not yet clearly identified for the latter group (which is the largest and includes children). Children living on a Hispanic-owned farm[3] face different occupational risks and perceive these risks differently than do adolescent farmworkers, recently arrived from Mexico or Central America, who are farm laborers. In both instances, children work along with their farmworker parents to contribute to their family income, but different risks may be involved for migrant farmworker children and for children on family owned farms. In either case, because children are physically immature, they are more susceptible than adults to work-related injury and disease.

Agricultural work poses exceptional injury risk factors for children (Castillo et al, 1998; Castillo et al, 1999; Villarejo and Baron, 1999). Repetitive lifting, bending, stooping, and movements of the hands and wrists in hand-intensive fieldwork expose children to additional hazards (Mobed et al., 1992). Trying to maintain a rapid work pace may place children at risk of future disability (Strong and Maralani, 1998). Being unaccustomed to the work increases the risk of musculoskeletal disorders for new workers, especially for children (Hakkanen et al., 2001). The effects of learning new skills under an unfamiliar environment need to be considered in Spanish-speaking children's health. Children's responses to the agricultural hazards and strenuous circumstances may predict their future disability, but we have no data on this.

In addition to the impact of physical labor, long-term effects of exposure to herbicides, pesticides and other chemicals on the well-being of farmworker children have not been described

---

[1]The term "Hispanic" includes all farm operators of Spanish, Hispanic, or Latino origin regardless of race.

[2]"Farmworker" refers to hired persons who leave their homes and migrate to work in agriculture in one or more states. "Farmworker" also includes people who are hired to work locally in seasonal agricultural jobs but do not leave their permanent residence. The term "adolescent" includes children 12 through 17 years old.

[3]The census of agriculture defines a farm as any place from which $1,000 or more of agricultural products were produced or sold during the reference year.

or assessed. Biological and toxic chemicals may have different effects at different stages of physiological development and may be more damaging to children than to adults with similar exposures.

Risks faced by Spanish-speaking children go beyond agricultural hazards. Motor vehicle crashes are the leading cause of injury-related deaths for Hispanics; poisonings are second (CDC, 1998). Hispanic children are at higher risk than non-Hispanic whites for injuries resulting from violence. Hispanics' lower education levels and higher poverty levels, along with the family disruption and weak intergenerational ties caused by migration, have been shown to increase risk for violent behavior.

The number of Spanish-speaking children working in agriculture is large and the likelihood of work-related injury and disease is high (Castillo et al., 1999; Kebebew, 1998; Stallones and Switzer, 1999). Some health and safety preventive measures have been developed to protect children on the traditional U.S. family farm (Lee and Marlenga, 1999; Lee, et al., 2002; NCASH, 1988; NCCAIP, 1996;). However, Spanish-speaking children in agriculture merit interventions developed and targeted especially at them: one size does not fit all. All efforts require thorough evaluation from the perspective of culture and the Spanish language. The diverse needs of Spanish-speaking children and their different environments must be considered. The integration of socio-cultural, developmental, and language circumstances will be a challenge for health and safety professionals. Occupational health risk interventions for Spanish-speaking children can begin with multiparty translations of existing intervention models and the targeting of suitable resources toward this new workforce. For the promotion of health and safety information to succeed in this multicultural and multilingual agricultural industry, the art and science of communication of health risks must be transformed to meet the special needs of children in agriculture.

Communication of occupational risks to agricultural employers, parents, and especially Spanish-speaking children requires concentrated efforts to convey culturally appropriate messages to meet the needs of each target group (Finau, 2000). Adequate risk communications engage a tripartite interaction between the sender (health and safety professional), receiver (employer, parent, child) and the medium (hands on, written, visual) (Finau, 2000). The ultimate goal of health and safety professionals must be to provide the optimal environment for all these children to become healthy adults.

AGRICULTURE: A DANGEROUS INDUSTRY FOR CHILDREN

The National Institute for Occupational Safety and Health (NIOSH) reports more than 2.25 million full-time workers employed in agriculture. If unpaid farm workers and family members 14 and older are included, nearly 4.5 million persons work in agriculture (CDC, 2001). Even though the U.S. agricultural sector provides an increasing and affordable supply of food and fiber, agriculture continues to have the second highest fatal occupational injury and serious non-fatal injury rates for U.S. workers (DOL, 2000a,c). In 2000 the Bureau of Labor Statistics reported that nearly 6,000 workers were killed on the job from traumatic injuries and more than 6.3 million suffered other injuries or illnesses (Caswell et al., 2001; DOL, 2000a). The national cost of injury has been estimated to be more than $50 billion annually (Castillo et al., 1998; DOL, 2000a). Costs for job-related injuries and illnesses are higher than those for AIDS and Alzheimer's disease and are on a par with those for cancer and circulatory disease, the two greatest causes of mortality (CDC, 2001).

The exact number of children working by industry sector is unknown due to diversity in definitions and seasonal variations, particularly in agriculture. Data show that over half of children have held some kind of job in industry by the age of 14 (DOL, 2000c). There are approximately 1 million children (under 15 years of age) who reside in farm operator households, while another 800,000 children live in households headed by hired farmworkers (CDC, 2001). During 1996-98, 2.9 million children (aged 15-17) worked during the school months, and 4 million during the summer months (DOL, 2001c). In 1996 working children were estimated to be 7.6 percent (262,000 children 16-19 years of age) of the workforce in the agriculture, forestry, and fishing industries (DOL, 2000a,b).

Agriculture employs most of the world's working children (DOL, 1995). All over the world children labor in occupations and industries that are dangerous or hazardous (CDC, 1995; DOL, 1995). They are exposed to hazards and often perform tasks that are beyond their physical capacity, such as lifting and carrying heavy loads and handling dangerous tools and equipment. Children encounter sharp and unwieldy tools, bites from insects and snakes. They are transported in unsafe vehicles and are regularly exposed to toxic chemical substances (fertilizers, and pesticides). Children often work without protective clothing and under extreme temperatures (Grainger, 1993).

Children working in United States fields hold some of the lowest-paying jobs in the country (DOL, 2000c). They are paid less than the already low wages of their adult counterparts. The decision to have a child work is based on short-term economic considerations to the potential detriment of the child's long-term health and development (DOL, 2000b). Because different laws govern employment of children in agriculture, most remain invisible to health and safety professionals. Many children in agriculture are not included in the population at risk, since only legally working children are recorded (McLaurin, 2000). This evidence suggests a need to redesign institutions to focus on children's rights, to align the interests of their parents, the agricultural industry, labor, and professional organizations in order to provide the best environment to prevent harm.

Agricultural injuries to children are documented as a public health problem (Castillo et al., 1998; NCCAIP, 1996;). The rates of agricultural work-related fatalities in children are disturbing (Castillo et al., 1999). Child fatality rates are 2.4 times greater than the overall work-related fatality rate for the United States across all ages and all industries (Castillo et al., 1999; DOL, 2000a;). While this statistic is alarming, it may be only the tip of the iceberg. Most estimates indicate that occupational diseases account for far more fatalities than occupational injuries. Precise enumeration of occupational disease fatalities is difficult due to paucity of data, owing to an under-diagnosis of occupational diseases and the inadequacy of surveillance systems (Herbert and Landrigan, 2000).

Spanish-speaking children working in agriculture are not a homogeneous group. There are children working on Hispanic family farms. There are an uncertain number of "invisible" children who migrate along with their parents or family members to work on U.S. farms. Spanish-speaking adolescent farmworkers (14-17 years old) are estimated to be 7 percent of the 2 million migrant and seasonal workforce (DOL, 2000b). And, there are other Spanish-speaking children not yet included in these agricultural statistics. For example, significant numbers of children work in the fishing industry. Children dive for fish, work on fishing platforms and boats, collect shellfish and shrimp larvae, peel shrimp, and clean fish. In performing these tasks they often spend long hours in the water and face such hazards as drowning, skin diseases, and attack by dangerous fish. They also risk injury from the tools used for cutting and cleaning fish and seafood. The nature of their work exposes them and their families to an exceptionally high risk of injury and illness (Grainger, 1993).

The effects of children's earnings and workload on their well-being, including mental health, have not been studied in detail. Not all work performed by children may be detrimental or exploitative (DOL, 1998). Children's work can be viewed as beneficial to the child, family, and society in general. Children working in agriculture as part of a family unit may reduce their parents' workloads, particularly if earnings are based on weight or piece. They may substitute for parents on domestic tasks, taking care of younger siblings or other family members, while parents are working in the fields for extended hours. Family dependence on children's earnings may contribute to keeping those same children in poverty, since they are often deprived of educational and other opportunities (DOL, 1995; Otis et al., 2001).

Children's lack of developmental skills may increase the probability of agricultural injuries. Child workers are also more susceptible than adults to work hazards due to anatomical differences. The higher vulnerability of children to such environmental stressors as temperature changes, ionizing radiation, and chemicals place them at higher risk than adults for carcinogens, occupational injuries, permanent disabilities, and diseases (Bequele and Meyers, 1995).

CHILDREN FROM HISPANIC FAMILY FARMS

Hispanics are traditional agriculturalists. Since 1978 the number of farms operated by persons of Hispanic origin increased 58 percent while the total number of farms decreased 15 percent. In 1997 farms operated by persons of Hispanic origin sold commodities worth $3.3 billion, a total of 2 percent of total U.S. crop sold that year. In 1997 the value of vegetables, nuts, berries, and fruits was 22 percent of U.S. crops sold. In contrast, for the same year vegetables, nuts, berries, and fruits accounted for more than 50 percent of crops sold from operators of Hispanic origin. During the last few decades an increased demand for farmworkers to conduct labor-intensive agriculture has been tied to increasing U.S. production of fruits and vegetables (Villarejo and Runsten, 1993). Health outcomes for Hispanics working as hired farmworkers or operators have not yet been elucidated. Operators and hired farmworkers of Hispanic origin have been described as being at significantly increased risk of death from head injuries compared with non-Hispanics in the same job categories (Stallones and Sweitzer, 1999).

Estimated injury rates for family workers (some of which are children) on all farms are higher (8.3 injuries per 200,000 hours) than rates for hired workers (4.9 injuries per 200,000 hours) (DOL, 2000c). Occupational risks for children of Hispanic operators are yet to be characterized. Hispanic family farms may provide working opportunities for children on a seasonal basis: full-time during planting and harvest seasons and part-time during the remainder of the year. Spanish-speaking children working on family farms face different situations than children (emancipated minors) working seasonally as hired farmworkers or migrant workers. For example, we do not know what concerns regarding preventing an injury, or disease process, would face a parent who previously was a migrant farmworker while assigning tasks for their children at their own farm?

SPANISH-SPEAKING MIGRANT CHILDREN AND ADOLESCENT FARMWORKERS

The agricultural industry employs about 2.5 million farmworkers. Hired farmworkers,[4] some of whom are adolescents, are four times more likely to die from an occupational injury than to non-farm workers (Crandall et al., 1997). According to the Traumatic Injury Surveillance of Farmers Survey project, Hispanics account for the second highest number of injuries overall (16.8 percent). Hispanics accounted for 51.9 percent of the estimated 59,888 injuries among hired farmworkers (CDC, 2001).

The National Agricultural Worker Survey (NAWS) is conducted for agriculture crop operations (DOL, 2000b). According to NAWS, more than 70 percent of the hired crop farmworkers are foreign born (from Mexico and Central America) (Gabbard, et al 1999). Spanish is the primary language of more than two-thirds of these foreign born farmworkers (94 percent of whom are Mexican nationals) (Villarejo and Baron, 1999). One-third of these foreign born

---

[4]"Hired farmworker" is defined in the U.S. census as an occupation employed to perform on-farm tasks for the purpose of producing an agricultural commodity for sale.

farmworkers do not have authorization to work in the U.S.; they have low education levels and many live in poverty. Mexicanization of the U.S. hired farm workforce, during the last few decades, has several implications for occupational health and safety. An estimated 126,000 farmworker children (14 to 17 years old) performed farm work each year from 1993 to 1997 (Gabbard et al., 1999). Children migrating along with their farmworker parents or relatives face the same hazards as their parents (DOL, 2000b; Villarejo and Baron, 1999). The NAWS data shows the number of children working in agriculture along with their parents has decreased over the last several years. In addition most children of farmworkers are very young (DOL, 2000b). These younger children can be exposed as bystanders to the same work environment and hazards in the fields as their adult family members (DOL, 2000c). Migrant farmworker children are found mostly at vegetable and fruit operations. Vegetable and fruit operations yield the third highest reported number of lost workdays as a result of injuries (Castillo et al., 1999).

Nearly two-thirds of all farmworkers live below the poverty threshold, causing difficulty in accessing health care and education. Under-reporting of health conditions is significant as a result of limited access to health services. A small number (10 to 11 percent) of farmworkers receive health insurance through their employers (USDA, 1992; DOL, 2000b;). As a result, most farmworkers and their dependents do not seek regular health care services and look for services only in emergencies. Several studies have emphasized that farmworker families have different cultural concepts of health, injury, and disease (Meade and Calvo, 2001; Villarejo and Baron, 1999; USDA, 1992).

Crop agriculture is increasingly the workplace of foreign-born children (CDC, 1995; DOL, 2000b). Adolescent farmworkers consists of unmarried, young males and emancipated minors traveling long distances to obtain temporary agricultural employment (Vela-Acosta and Lee, 2001). Most adolescent farmworkers come from Mexico as emancipated minors to work while living away from their home; they are recent arrivals who are not authorized to work in the U.S. (Gabbard et al., 1999). Their working and living conditions include an exceptional combination of hazards, made worse because they are already a disadvantaged group (Vela-Acosta and Lee, 2001). In addition to exposure to agricultural hazards, all adolescent farmworker are facing additional challenges such as extreme poverty; variable working and living conditions; loss of educational opportunities; separation from their familiar environment; and lack of parental supervision (Castillo et al., 1999). Because of the need to focus first on survival, adolescent farmworkers can jeopardize their own potential. Seasonal agricultural work often conflicts with children's school attendance. Children frequently miss classes, and some are even forced to give up years of education. On the job children lack medical care, overtime pay, rest breaks, rest days, and formal education. Defenseless migrating children working in agriculture face numerous additional hazards, such as transportation in unsafe vehicles, risk of violence, and physical assault at unsecure locations (DOL, 1995).

EFFORTS ADDRESSING OCCUPATIONAL HEALTH AND SAFETY FOR CHILDREN IN AGRICULTURE

In 1995 NIOSH organized a national advisory group consisting of clinical, research policy experts to provide advice and direction on occupational health surveillance of adult farmworkers. The major priorities noted by this group are shown in Box 1. Note that children of farmworker families and adolescent farmworkers are working and living under the same conditions as adult farmworkers. All priorities outlined for farmworkers need to be considered in realistic ways for children as well.

Box 1
Workgroup Reported Priorities for Surveillance and Research of the Occupational Safety and Health of Hired Farmworkers

1. Surveillance
Ergonomics, pesticides, traumatic injuries, water quality, respiratory, dermal, eye, infectious diseases, cancer and mental health

2. Research
Pesticides, ergonomic, injuries, effect of protective measures, cancer, comparison population and mental health

3. Project Design Methods
Regional and local studies, national and international studies, special populations

Source: CDC, 2001.

In response to children being injured while living, working, or visiting agricultural work environments (primarily traditional English-speaking farms) the National Committee for Childhood Agricultural Injury Prevention reached consensus on research, education, policy, and other interventions aimed at the reduction of agricultural injuries among children; in October 1996 NIOSH announced support of these recommendations (see Box 2). As a result North American Guidelines for Children's Agricultural Tasks (NAGCAT), supported by NIOSH, developed non-regulatory guidelines for children working under parental supervision (Lee and Marlenga, 1999). Overall, preventive initiatives to protect children working in agriculture have not been directed at issues affecting Spanish-speaking children, particularly migrant children. Work has focused only on children who reside on family farms, and for the most part only on injury prevention (Lee and Marlenga, 1999; Lee et al., 2002; NCASH, 1988; NCCAIP, 1996).

Health care for migrant children demands special consideration by health professionals. In 2000 the American Academy of Pediatrics joined efforts with the Migrant Clinicians Network and completed guidelines for the care of migrant farmworker children (McLaurin, 2000). This manual assists clinicians in becoming familiar with the risks and characteristics of migrant farmworker children. Some guidelines from the manual are clearly linked to the occupational health of Spanish-speaking children (see Table 1) (McLaurin, 2000). Health care experts serving Spanish-speaking children are urged to work to provide a hazard-free environment at worksites, home and schools.

The National Adolescent Farmworker Occupational Health and Safety Advisory Committee (NAFOHSAC) (see Box 3) is the first national effort to address adolescent farmworkers as a population at risk (Vela-Acosta and Lee, 2001). The multidisciplinary committee consisted of farmworker parents, adolescent farmworkers, researchers, non-government organizations, health care providers, federal and state agency representatives, and agricultural employers. The committee gathered information from published literature, external sources, committee discussions, focus groups, and a committee workshop. The report review involved input from committee members and peer reviewers across the United States (Vela-Acosta and Lee, 2001). NAFOHSAC has the potential to create a foundation for improving the working conditions of adolescent farmworkers (Lee et al., 2002).

Box 2
Recommendations from the National Committee for Childhood Agricultural Injury Prevention

1. Establish and maintain a national system for childhood agricultural injury prevention.
2. Ensure that childhood agricultural injury prevention is supported with sufficient funding and cooperation from the public and private sectors.
3. Establish guidelines for children's and adolescent's work in the industry of agriculture.
4. Ensure that the public is aware of general childhood agricultural safety and health issues.
5. Establish and maintain a comprehensive national database of fatal and non-fatal agricultural injuries.
6. Conduct research on costs, risk factors, and consequences associated with children and adolescent who participate in agricultural work.
7. Use systematic evaluations to ensure that educational materials and methods targeted toward childhood agricultural safety and health have demonstrated positive results.
8. Ensure that farm and ranch owner/operators and parents understand relevant agricultural safety and health issues that pertain to children and adolescents.
9. Ensure that rural safety and health professionals understand the issues relevant to children and adolescents exposed to agricultural hazards.
10. Influence adult behaviors that affect protection of children and adolescents through the use of incentives and adoption of voluntary guidelines.
11. Provide a protective and supportive environment for children exposed as bystanders to agricultural hazards.

Source: NCCAIP, 1996.

TABLE 1 Recommendations for Pediatricians about the Health Care of Migrant Farmworker Children

Health Care Guidelines	Policies Affecting Migrant Farmworkers
1. Well-Child Visits	1. Fair Labor Standards Act
2. School Readiness	2. Migrant Health Programs
3. Immunizations	3. Medicaid
4. Adolescent Care	4. Vaccines for Children
5. Oral Health	5. Supplemental Nutrition Program for Women, Infants, and Children (WIC)
6. Environmental Concerns: Lead, pesticides, groundwater contamination	6. Foster Care, Adoption and Children with Special Needs
7. Child Maltreatment	7. State Children's Health Insurance Program (SCHIP)
8. Injuries: Machinery, fatal injuries, animal-related, falls Violence, legal, and socio-cultural factors	8. School Entry
9. Infectious Diseases: Tuberculosis, acquired immunodeficiency virus (AIDS), parasitic infestation, protozoan infections, zoonosis (Leshmaniasis), malaria, congenital syphilis	

Source: McLaurin, 2000.

Box 3
Goals from the National Adolescent Farmworker Occupational Health and Safety Advisory Committee

Goal I: Identify profiles of hired adolescent farmworkers employed in production agriculture across the United States.

1. The Department of Labor (DOL) and the National Institute for Occupational Safety and Health (NIOSH), with the cooperation of migrant non-profit organizations, state health departments, and the Migrant Clinicians Network (MCN), should be funded to develop and maintain a regional ongoing database of occupation demographics.
2. Wage and hour divisions at state levels, in coordination with migrant agencies, DOL, NIOSH, and the U.S. Department of Agriculture (USDA) should coordinate efforts to assess occupational risks at adolescent farmworkers' worksites and housing locations, which often are separate. Worksites where adolescent farmworkers are employed should provide facilities and resources that address adolescents' needs, including personal hygiene, emergency contact information, telephone communications, and adult supervision.

Goal II: Identify occupational risks that are potentially unique and specific to hired adolescent farmworkers.

3. Congress should allocate funds to the Centers for Disease Control and Prevention (CDC), designating NIOSH to plan, implement, and evaluate intramural and extramural research to promote best work practices and to improve the health and safety of hired adolescent farmworkers. NIOSH can convene a group of agricultural safety specialists, agricultural producers, and occupational health care providers who work with hired adolescent farmworkers to evaluate progress to improve their working conditions. Information gathered by these experts on ways to minimize risk factors should be disseminated to health professionals, agricultural employers, and others serving adolescent farmworkers to help them understand agricultural risks by commodity groups.
4. Congress should allocate funds to the CDC, designating NIOSH as the leading agency along with the DOL and the USDA, to establish data collection methodologies that will identify major sources of occupational risks, disease, and injury among adolescent farmworkers by crop, region, and type of employer. Comprehensive analysis of exposure to work hazards should integrate environmental conditions, ergonomic hazards, and physiological factors for adolescents. Research efforts need to be targeted to regions where adolescent farmworkers are concentrated.
5. The Environmental Protection Agency (EPA), DOL, migrant health professionals, agricultural health and safety professionals, agricultural employers, and others should provide language- and culture-appropriate access (e.g., toll-free telephone) for adolescent farmworkers to express their work questions and concerns about occupational hazards.

Goal III: Plan, implement, and evaluate interventions to eliminate or minimize occupational health and safety risks of hired adolescent farmworkers.

6. Congress should fund the NIOSH Agricultural Centers and the USDA Cooperative Extension Services, through the Risk Management Education Division (RMED), to support agricultural employers' associations in the establishment of systems to educate agricultural employers, supervisors of adolescent farmworkers, and farm labor contractors about adolescent farmworkers' occupational risks and prevention strategies.
7. Agricultural employers' organizations and others should facilitate discussions among members and researchers to promote best work practices helping agricultural employers to identify practical solutions for occupational hazards among adolescent farmworkers. They can provide time during regional, state, and national meetings for major presentations and group discussions to report innovative and effective interventions.

8. NIOSH should be funded to provide agricultural employer-targeted evaluation studies to determine the cost-effectiveness of intervention programs aimed at preventing occupational disease and injury among adolescent farmworkers.
9. The American Academy of Pediatrics (AAP), North American Agromedicine Consortium, NIOSH, and other relevant organizations should provide advanced training for health professionals on occupational health and safety conditions in agriculture, with a special focus on adolescent farmworkers.
10. NIOSH should evaluate the impact of the agricultural media in enhancing public awareness to effectively promote best work practices involving adolescent farmworkers among agricultural employers.
11. A coordinated, regional approach should be established to address adolescent farmworkers' occupational health and safety needs. Funding should be available to universities, NIOSH Agricultural Centers, and others to provide adolescent farmworker educational health and safety interventions. Those actions will (1) maximize collaborative efforts with current initiatives; (2) develop language and culturally appropriate materials; (3) assess adolescent farmworkers' safety risk perceptions; (4) account for cultural beliefs about safety practices; and (5) be tailored specifically to adolescent learning needs and not just a part of a general session for all workers.
12. Congress should create a Farmworker Adolescent Network (FAN) and designate the DOL, the USDA, NIOSH, and the EPA to lead and coordinate a multi-organization network for addressing adolescent farmworkers' occupational health and safety research and program activities. Organizations representing agricultural employers, migrant health professionals, migrant advocates, adolescent farmworkers, and agricultural safety professionals should be represented in this new FAN. Joint venture funds from public and private sectors should be encouraged to support FAN to fund regional and national initiatives, including data collection, research, training, and innovative prevention programs based on significant research findings from convened groups and FAN.

Source: Vela-Acosta and Lee, 2001.

The Childhood Agricultural Injury Prevention Progress Report and Updated National Plan (NCCAIP, 1996) has reviewed and updated the progress by the National Committee for Childhood Agricultural Injury Prevention (see Box 4) (Lee et al., 2002). The report emphasizes that education is not enough to deal with childhood injuries. The update recommends a multifaceted approach that incorporates systematic evaluations in addition to empowerment of agricultural key players (farm families, rural schools, farm and community groups, agribusinesses, and agricultural trainers). Concerns expressed in the report, include the lack of efforts to address children at risk in agriculture other than those on traditional family farms. In addition, the adoption of programs by agricultural employers needs to be cost effective and advantageous to them if they are to be used.

FUTURE DIRECTIONS FOR COMMUNICATING RISKS TO SPANISH-SPEAKING CHILDREN

Many factors need to be considered in reducing risks for Spanish-speaking children who work in the agricultural industry (see Box 5). Children's basic necessities (nutrition, education, rest, adult guidance, support for development, and recreation) in addition to agricultural work demands, financial, and socio-cultural needs should be considered. Interventions for Spanish-speaking children should be matched to their unique requirements. NIOSH expertise in agriculture can provide the necessary leadership to improve the health and safety of Spanish-speaking children.

Box 4
Goals of the Childhood Agricultural Injury Prevention Progress Report and Updated National Plan from the 2001 Summit

1. Adults will ensure that young children and non-working youth can grow, play, learn, and rest in protective environments that are free of agricultural hazards.
2. Young workers will receive agricultural safety training, guidance, personal protective equipment and adult supervision based on child development principles.
3. A strong public and private infrastructure will be maintained to ensure the vision, leadership, and national commitment necessary to prevent agricultural injuries.

Source: Lee et al., 2002.

Appropriate interventions must be based on collective efforts by government, health and safety professionals, agricultural employers, parents, and educators (CDC, 2001). Cost-effective measures need to be implemented to promote the adoption of best work practices for Spanish-speaking children.

How can a parent, educator, or agricultural employer affect the occupational risks and health outcomes at the worksites for Spanish-speaking children? Interventions have been concentrated on educational materials, assessments, and evaluations for English-speaking children who reside on family farms (NCASH, 1998; NCCAIP, 1996; Lee and Marlenga, 1999; Lee et al., 2002).

Safety guidelines, which can be used as a starting point, are available for children performing agricultural tasks (Lee and Marlenga, 1999). Spanish-speaking children are a unique population at risk, and careful evaluation for the appropriateness of these guidelines to their situation needs to be considered. Are those resources suitable for parents who supervise their Spanish-speaking children? How can a parent, educator, or agricultural employer prevent adverse health outcomes in the work experiences for Spanish-speaking children? Agricultural tasks performed by Spanish-speaking children, particularly by migrant and adolescent children, warrant serious consideration by NIOSH.

Communication of occupational risks must be tailored for each audience (see Box 5). Lack of understanding of the importance of culture has been reported as a major obstacle in addressing health disparities (Kritek et al., 2002). Culturally appropriate program goals can be attained by integrating the unique perceptions of risk for Spanish-speaking children. While advertisements, written materials, and the media may be helpful, community efforts should include employers as partners in order to make materials accessible at sites where the young Spanish workers may receive the information (Ford et al., 2001).

As noted earlier, exposures to agricultural hazards may lead children to injury, disease, or permanent disability (Bartels et al., 2000; Kebebew, 1998; Rittichier and Bassett, 2001). Rarely in any occupational setting is there an exposure to a single agent. Rather, it is a mixture of hazards that may act antagonistically, synergistically, or additively. Children are exposed to work hazards which vary in agricultural workplaces, and hazards are changing as the result of new technologies. Once occupational risks are identified for each group of Spanish-speaking children by commodity and region, clear messages to address these risks need to be developed. Straightforward, practical messages in their own languages are the means to empower Spanish-speaking children about the best ways to protect themselves from physical (heat, noise, vibration), biological (infectious agents), and chemical hazards (see Box 5).

Risk factors for Spanish-speaking child populations are more complex than those traditionally seen in agriculture (see Boxes 3 and 5). Messages for the children need to be consistent with their reality. For example, farmworker populations, including their children, often

Box 5
Factors to Consider in Addressing and Communicating Occupational Risks to Spanish-Speaking Children

At-risk groups

- People living on Hispanic farms
- Traveling with migrant farmworkers:
- Bystanders
- Working
- Adolescent farmworkers
- Still 'invisible': fishing industry

Hazards

- Physical (transportation, noise, heat)
- Chemical (pesticides, insecticides, fertilizers, lead)
- Mechanical (motor vehicle, farm tools,
- Biological (sanitation, infectious agents, immune response)
- Psychosocial (ddevelopmental stages, cultural background, financial need, substance abuse)
- Ergonomic (appropriateness, body size)

Risk Factors

- Age
- Gender
- Language skills
- Education
- Cultural beliefs
- Perception of risk
- Financial need
- Family influence
- Work experience
- Adult supervision
- Geographical location
- Agricultural task
- Agricultural crop
- Migration
- Labor law
- Legal status
- Mental health
- Developmental status
- Agricultural contractor
- Housing

Audiences for Reducing Risk

- Hispanic farm families
- Migrant and seasonal farmworkers
- Adolescent farmworkers
- Community based organizations
- Government and federal organizations
- Health and safety professionals
- Agricultural employers
- Educators and trainers

Note: In this table, "hazard description" is not all-inclusive but is intended to highlight overlying effects and hazards of particular concern for children. These hazards are specific to the constantly changing agricultural work environment.

have no control over the provision of sanitary facilities in the fields or at labor camps. Lack of access to decontamination facilities exacerbate their exposure to chemical and biological hazards due to prolonged skin contact (EPA, 1993; Vela-Acosta, 1999a,b; Vela-Acosta et al., 2002).Carefully designed safety messages are ineffective for the adoption of safety behaviors when work and living environments do not support the suggested modified behavior (Vela-Acosta, 1999a,b; Vela-Acosta et al., 2002). Strategies to increase adequate environment and adoption of safe behaviors in agriculture must be cost effective for both workers and employers. Clear benefits need to be communicated to children, agricultural employers, parents, and educators.

Access to receptive audiences can start with community organizations, agricultural employer organizations, migrant organizations, and others reaching this population. Comprehensive, coordinated efforts are necessary to prevent agriculture-related injuries and diseases for all Spanish-speaking children and adolescents (see Boxes 3 and 5). Communicating benefits of safety and health provisions to each audience is very important. For example, operators of the 2.75 million farms and ranches need to embrace appropriate preventive innovations. Safety interventions often have a low rate of acceptance when benefits will occur in the future and are directed to protect renewable human resources while not directly increasing farm profits (Caswell et al., 2001).

Additional efforts at agricultural childhood initiatives must include an extraordinary population at risk: Spanish-speaking children. Health and safety professionals are obligated to provide better guidance to agricultural employers and parents after evaluating work exposures and work benefits for children in agriculture. The most effective methods for communicating occupational risks should be identified by researchers and taught to agricultural employers, educators, and parents. Clear risk communications are required to have a significant impact on changing the environment provided for children. Public health professionals and policy makers should be informed about the risks children face in their communities. All changes in communicating occupational risks must take effect through collaboration between government, industry, labor, and academia.

Each message being disseminated to a particular audience needs to be functional. Practical and theoretical reference materials should be emphasized. Research findings and new technology should be readily transferred from scientists to health and safety professionals, employers, educators, and parents to ensure reduction of occupational injuries and diseases resulting from exposures to each agricultural work environment. The first step in decreasing occupational risks is to adequately communicate the risk in the words of the population in need. Adequate interventions are needed to modify living and working farm environments to decrease disease and injuries and ensure healthy young adults.

Ideally, this work to improve safety needs to be regional and international. Multinational health and education outreach programs need to be developed and funded in order to serve the best interest of the health of all future working adults (Villarejo and Baron, 1999). Goals from previous childhood initiatives, relevant to Spanish-speaking children, such as recommended actions from the NAFOHSAC (particularly under Goal II and III, Box 3), and other relevant initiatives can set preliminary objectives to be supported by NIOSH throughout partnerships with universities, agricultural research, and health centers to minimize the occupational health and safety risks of all Spanish speaking working children (Lee and Marlenga, 1999; Lee et al., 2002; McLaurin, 2000; NCCAIP, 1996; Vela-Acosta and Lee, 2001). In addition, efforts are needed to eliminate risks for all working children regardless of their native language. Regional approaches are required to involve community organizations, private initiatives, and all interested parties to protect the future well-being of Spanish-speaking children.

REFERENCES

Bartels, S., B. Niederman, and T.R. Waters. 2000. Job hazards for musculoskeletal disorders for youth working on farms. J Ag. Safety and Health. 6(3):191-120.

Bequele, A., and W.E. Myers. 1995. First things first in child labor: Eliminating Work Detrimental to Children (Geneva: UNICEF and the International Labor Organization, 9).

Castillo, D., D. Hard, J. Myers, A. Pizatella, and N. Sout. 1998. A national childhood agricultural injury prevention initiative. J. Ag. Safety and Health. 1:183:191.

Castillo, D.N., N. Adekoya, and J.R. Myers. 1999. Fatal work-related injuries in the agricultural production and services sectors among youth in the United States. J Agromedicine, 6(3):27-41.

Caswell, M., K. Fuglie, C. Ingram, S. Jans, and C. Kascak. 2001. Adoption of Agricultural Production Practices. Lessons learned from the U.S. Department of Agricultural Area Studies Project. Economic Research Service, AER: 792. ERS Agricultural Economic Report No. 792:116.

CDC (Centers for Disease Control and Prevention). 1995. United States Department of Health and Human Services, National Institute for Occupational Safety and Health (1995). Workgroup on Priorities for Farmworker Occupational Health Surveillance and Research. New Directions in the Surveillance of Hired Farm Worker Health and Occupational Safety. Cincinnati, OH, NIOSH. Online. Available at www.cdc.gov/niosh/hfw-index.html#contents. Accessed May 16, 2002.

CDC (Central for Diseases Control and Prevention). 1998. Injury Fact book 2001-2002. Online. Available at http://www.cdc.gov. Accessed June 25, 2002.

CDC (Centers for Disease Control and Prevention). 2001. United States Department of Health and Human Services. National Institute for Occupational Safety and Health (May 2001). Injuries Among Farm Workers in the United States, 1995. Myers, J.R., Publication No. 2001-153. Online. Available at www.cdc.gov/niosh. Accessed June 19, 2002.

CDC (Centers for Disease Control and Prevention). 2001. United States Department of Health and Human Services. National Institute for Occupational Health and Safety Publications 2001-158. Online. Available at www.cdc.gov/niosh/2001-158.html. Accessed May 15, 2002.

Crandall, C.S., L.Fullerton, L. Olson, D.P. Skylar, and R. Zumwalt, R. 1997. Farm-related injury mortality in New Mexico, 1980-91. Accid. Anal and Prev. 29(2):257-261.

DOL (U.S. Department of Labor). 1995. By the Sweat and Toil of Children volume II. The Use of Child Labor in U.S. Agricultural Imports and Forced and Bonded Child Labor. A Report to the Committee on Appropriations, U.S. Congress, Department of Labor, Bureau of International Labor Affairs, Otero, J.F., Deputy Under Secretary for International Affairs: Child Labor in Commercial Agriculture.

DOL. 1998. Congressional-mandated report in the international child labor series of the Department of Labor's Bureau of International Labor Affairs. Departments of Labor, Health and Human Services, and Education, and Related Agencies Appropriations Bill of 1998, P.L., pp. 105-78.

DOL. 2000a. Bureau of Labor Statistics. Injuries and Fatality Data. Online. Available at : www.stats.bls.gov/iif/home.htm. Accessed May 17, 2002.

DOL. 2000b. Findings from the National Agricultural Workers Survey: 1997-1998. A demographic and employment profile of United States farmworkers. Research Report No. 8.

DOL. 2000c. Report of the Youth Labor Force. Online. Available at www.bls.gov. Accessed May 17, 2002.

EPA (Environmental Protection Agency). 1993. The Worker Protection Standard for Agricultural Pesticides, How to Comply. United States Environmental Protection Agency 735-B-93-001. Prevention, Pesticides, and Toxic Substances (H75506C). SSOP, Washington, DC 20402-9328.

Finau, S.A. 2000. Communicating health risks in the Pacific: scientific construct and cultural reality. Asia-Pacific Journal of Public Health. 12(2):90-97.

Ford, K., G. King, L. Nerenberg, and C. Rojo. 2001. Aids knowledge and risk behaviors among Midwest migrant farm workers. AIDS Education and Prevention. 13(6):551-560.

Gabbard, S., D. Carroll, S. Baron, and A. Steege. 1999. Teens in Crop Agriculture. Paper prepared for the National Adolescent Farmworker Occupational Health and Safety Advisory Committee. Washington, DC: US Department of Labor.

Grainger, C.R. 1993. Hazards of commercial fishing. World Health Forum. 14(3):313-315.

Hakkanen, M., E. Viikari-Juntura, and R Martikainen. 2001. Job experience, work load, and risk of musculoskeletal disorders. Occupational and Environmental Medicine. 58(2):129-135.

Herbert, R., and P.J. Landrigan. 2000. Work-related death: A continuing epidemic. American Journal of Public Health. 90(4):541-545.

Kebebew, A. 1998. Statistics on Working Children and Hazardous Child Labour in Brief. International Labour Organization (ILO) Office, Geneva (eds). First Published 1997, revised April 1998.

Kritek, P.B., M. Hargraves, E. Cuellar, F. Dallor, D.M. Gauthier, C.A. Holland, C. Ilkiw, J.W. Swanson, and R. Swanson. 2002. Eliminating health disparities among minority women: A report on conference workshop process and outcomes. American Journal of Public Health. 92 (4):580-587.

Lee, B. and B. Marlenga. 1999. Professional Resource Manual: North American Guidelines for Children's Agricultural Tasks. Marshfield, WI: Marshfield Clinic. Online. Available at www.nagcat.org or http://www.gemplers.com. Accessed May 25, 2002.

Lee, B., S. Gallagher, B. Marlenga, and D. Hard. 2002. Childhood Agricultural Prevention: Progress Report and Updated National Action Plan for the 2001 Summit. Marshfield, WI. Marshfield Clinic. Online. Available at www.research.marshfieldclinic.org/children/. Accessed May 17, 2002.

McLaurin, J.A. 2000. The Practice of Contextual and Integrative Child Care: New Opportunities. in Health Care for Children of Migrant Farmworkers. American Academy of Pediatrics (eds.).

Meade, C.D., and A. Calvo. 2001. Developing community-academic partnerships to enhance breast health among rural and Hispanic migrant and seasonal farmworker women. Oncology Nursing Forum. 28(10):1577-1584.

Mobed, K., E. B. Gold, and M. B. Schenker. 1992. Occupational Health Problems Among Migrant and Seasonal Farm Workers. In Cross Cultural Medicine: A Decade Latter (Special Edition). West J. Med 157:367-373.

NCASH (National Coalition for Agricultural Safety and Health). 1988. Agriculture at risk: A report to the nation. Iowa City, IA (eds.), The University of Iowa, pp. 1-67.

NCCAIP (National Committee for Childhood Agricultural Injury Prevention). 1996. National action plan, children and agriculture: Opportunities for safety and health. Marshfield, WI: Marshfield Clinic.

Otis, J., E. M. Pasztor, and E. J. McFadden. 2001. Child labor: a forgotten focus for child welfare. Child Welfare. 80(5):611-622.

Rittichier, K.K., and K.E. Bassett. 2001. Metal lawn and garden edging: the hidden knife? Ped Emerg Care 17(1):28-31.

Stallones, L., and K. Sweitzer. 1999. Fatal Head Injuries Among Hispanic Farmworkers and Farmers in Colorado, 1983-1992. Journal of Agricultural Safety and Health, 5(2):201-205.

Strong, M.F., and V. J. Maralani. 1998. in Farmworkers and Disability: Results of a National Survey. Berkeley, California, Berkeley Planning Associates (eds.).

USDA (U.S. Department of Agriculture) 1992. National Agricultural Statistics Service Farm Labor Report. Washington D.C., USDA.

Vela-Acosta, M.S. 1999a. Evaluation of a Risk Reduction Program Among Migrant and Seasonal Farmworkers in Colorado. Ph.D. Dissertation. Environmental Health Department. Colorado State University, Fort Collins, CO 80523.

Vela-Acosta, M.S. 1999b. Assessing The Effectiveness Of A Pesticide Training Program Among Migrant Farm Workers. High Plains Intermountain Center for Agricultural Health and Safety, Colorado State University, Fort Collins, CO American Industrial Hygiene Conference and Exposition. Online. Available at www.aiha.org/conf.html. Accessed on May16, 2002.

Vela-Acosta, M.S., and B. Lee. 2001. Migrant and Seasonal Hired Adolescent Farmworkers: A Plan to Improve Working Conditions. Marshfield, WI: Marshfield, Clinic. Online. Available at www.marshfieldclinic.org/nfmc/farmworkersreport/randr.stm. Accessed on April 29, 2002.

Vela-Acosta, M.S., P. Bigelow, and R. Buchan. 2002. Assessment of occupational health and safety risks of farmworkers in Colorado. American Journal of Industrial Medicine. 42(S2):19-27.

Villarejo, D. and S.L. Baron. 1999. The occupational health status of hired farm workers. Occupational Medicine. 14(3):613-635.

Villarejo, D., and D. Runsten. 1993. California's Agricultural Dilemma: Higher Production and Lower Wages. Davis, CA, California Institute for Rural Studies, (eds.) pp.1-48.

Appendix H

Occupational Health Among Latino Workers: A Needs Assessment and Recommended Interventions

Rafael Moure-Eraso, Ph.D., C.I.H.,
and George Friedman-Jimenez, M.D. M.P.H.

Note: An extended version of this material appeared as *Chapter 12: Occupational Health Among Latino Workers in the Urban Setting. In: M. Aguirre-Molina, C.W. Molina, and R.E. Zambrana (eds.). Health Issues in the Latino Community. San Francisco; Jossey-Bass 2001.* This manuscript has been updated to 2002.

INTRODUCTION

The occupational health of Latino workers in the United States is increasingly being recognized as an important area for study as well as for public health and clinical intervention. From a public health perspective this is an important issue to address for several reasons. The Latino population in the United States is sizeable (see Table 1) and growing rapidly, especially in urban areas. Work-related diseases cause substantial morbidity and mortality and are amenable to public health primary prevention interventions, such as the elimination or reduction of the exposures that cause them. Secondary prevention interventions in the form of surveillance and clinical services are also an essential part of the public health approach to occupational health. Although occupational diseases can affect members of all racial and ethnic groups and socioeconomic classes, available evidence suggests that Latino workers, along with other minority workers as well as low-income workers, are at higher risk for occupational disease than other workers in the general population. This excess risk is probably due to over representation of Latino workers in the more hazardous occupations and industries.

The resources allocated by the federal government for workers' health are modest. The National Institute for Occupational Safety and Health (NIOSH), Occupational Safety and Health Administration (OSHA), and Bureau of Labor Statistics (BLS) have budgets that are smaller by one order of magnitude than the budgets allocated to U.S. federal agencies with comparable missions, such as the Environmental Protection Agency (EPA), and the Food and Drug Administration (FDA). These allocations have consistently proved inadequate to serve the needs of the U.S. workforce in general. Consequently, the occupational health needs of minority workers (e.g., African Americans and Latinos) have been largely unmet with regard to targeted surveillance or primary prevention interventions.

Primary prevention is achieved by the elimination or substantial reduction of risk factors that are known to cause workplace death and disease. These interventions can only be successful with the full participation and cooperation of the groups affected by the hazards: workers and companies. A first step is to quantify the dimensions of the problem through hazard surveillance, or collection of systematic data on the prevalence of workplace hazards and populations at risk, followed by specific recommendations for engineering interventions at the point of production to control risk factors. Hazard surveillance should precede disease surveillance for the purposes of primary prevention.

Secondary prevention involves the early medical diagnosis and treatment of injury or illness that is successful in achieving recovery and return to work. The data on specific types of morbidity caused by the work environment in the Latino population are very limited, and many Latino workers lack access to available clinical occupational health services. As a result of this lack of epidemiological and surveillance data, programs providing clinical occupational health services to Latino working populations have not been developed adequately. This cycle of "no services → no data → no services" can be broken by (1) simultaneously developing epidemiological research and surveillance methods that will effectively include Latino workers and (2) by providing clinical occupational health services accessible by and targeted to Latino working populations.

A related topic of growing interest in public health for the Latino population is Environmental Justice (sometimes considered from the perspective of "environmental "injustice"). Environmental justice has focused on the observation that air polluting entities, hazardous worksites, hazardous waste dumps and other sources of environmental pollution are likely to be sited in close proximity to communities of color (Frumkin et al., 1999).

There are no systematic, reliable sources of data on occupational diseases in the U.S. working population (Herbert and Landrigan, 2000). Recent peer-reviewed estimates of occupational morbidity and mortality experience for the general U.S. population (1992), not differentiated by race, are substantial (Leigh et al., 1997). Using the best available denominator data for the U.S. working population, we developed an estimate of the number of occupational disease deaths and new cases among Latino workers in the United States. Next, using U.S. and New York City aggregate data, we see that Latino workers are disproportionately employed in the more hazardous occupational categories and under-represented in the less hazardous categories.

MINORITIES AS A PERCENTAGE OF THE U.S. WORKING POPULATION

The U.S. Census Bureau and the BLS of the U.S. Department of Labor publish population statistics for each year based on population projections from the 1990 census (U.S. Census Bureau, 2000). U.S. census data for 1998 and current statistics (1998) from the U.S. Department of Labor on U.S. occupational injury and illness experience and characteristics provided the basis for Tables 1 through 9. The demographic estimates of the U.S. working population by race and ethnicity appear in Table 1. The U.S. government provides data in Tables 1 through 9 combining race, (e.g., white and black) with ethnicity (e.g., Hispanic). Although not equivalent, we will use Latino for Hispanic and African-American for black.

TABLE 1 1998-99 Racial and Ethnicity Distribution of U.S. Population and Civilian Workforce

Race/Ethnicity	U.S. Civilian Workforce over 16 years old (in thousands)	Percent	Total U.S. Population (in thousands)	Percent
Hispanic (Latino)	14,492	10	34,864	12.8
Black	15,334	11	31,355	11.5
White	113,475	81	224,650	84.0
Total	140,863	100	272,820	100

NOTE: Percentages are greater than 100 because Hispanics can also be classified as black or white.
SOURCE: Census Bureau (2000) and BLS (2001).

It is interesting to note that Latinos are 10 percent of the civilian workforce (older than 16 years old) but 12.8 percent of the total population. In contrast, the African American civilian workforce is 11 percent while their proportion in the total population is 11.5 percent. The proportions of Latinos in both columns appear to show a substantial number of young Latinos (<16 years) not included yet in the civilian workforce.

Officially, only 5 percent of the Latino population works in agriculture (data on employed U.S. civilian population older than 16 years (DOL, 2000). However, other data sources estimate 12 percent in agriculture (EPA, 1999). A distribution of the 95 percent to 88 percent Latinos employed in occupations not in the agricultural sector shows that more than 67 percent are in blue-collar, low-paying jobs (service, labor and support and sales) while whites hold 56 percent of the same jobs (see Table 2a).

TABLE 2a Distribution of Occupations of Employed U.S. Civilian Workers Classified As Hispanic, Black and White Over 16 Years Old in 1994 (in thousands)

	Hispanic		Black		White	
Occupation	Number	Percent	Number	Percent	Number	Percent
Professional and managerial	1,517	14.0	2,405	20.2	30,045	28.6
Service (household, protective, other)	2,131	19.8	2,890	23.8	13,207	12.6
Operators, fabricators and laborers	2,474	22.9	2,677	22.0	14,416	13.7
Sales, administration, and technical support	2,639	24.0	3,637	29.9	32,232	30.7
Other	2,082	19.7	537	4.4	15,253	14.5
TOTAL	10,788		12,146		105,190	

SOURCE: Modified from EPA (1999). Table 7-3.

The most dramatic difference between the two groups is the proportion of Latinos in professional and managerial occupations (white collar) that is half of the proportion of whites in the same category (14 percent Latino vs. 28 percent white). African Americans follow a similar pattern except that their proportion in the professional and managerial classes is higher (20.2 percent) than Latinos. Still the proportion of blue-collar workers among African Americans is the highest at 75 percent. A more recent evaluation of the highest percentage of Latinos in selected occupations appears on Tables 2b and 2c.

OCCUPATIONAL DISEASE MORBIDITY ESTIMATE

Early estimation of the magnitude of occupational diseases in the United States showed that numbers of new cases could exceed 390,000 per year (Ashford, 1976). More recent estimates show a range of 817,015 to 907,385 of new cases of occupational disease annually. The newer estimate counted first the occupational deaths distributed among four major disease categories: cancer, coronary heart disease, cerebrovascular disease, and pulmonary disease (Leigh et al., 1997) (see Table 3).

To this number the authors added cases of non-fatal occupational disease reported by the U.S. Department of Labor's, Annual Survey of Occupational Illness in 1992, as well as the number of cases reported in the same year by public employees (identified as non-classified occupational disease in Table 3) (Leigh et al., 1997). Because the great majority of cases of occupational disease are not diagnosed as occupational and never become known to the surveillance mechanisms that currently exist, Leigh and colleagues used several indirect approaches to judge the magnitude and severity of the problem. However, the estimate of work-related musculoskeletal disorders, in

TABLE 2b BLS Labor Force Statistics of Selected Employed Persons by Detailed Occupation and by Race for 2001

Occupation Non-Agricultural	Total (thousands)	Hispanics %	Blacks %
Total working population	135,073	10.9	11.3
Professionals	41,894	5.1	8.3
Managers	5,139	5.7	9.7
Services	18,359	16.3	17.9
Household services	239	32.8	12.1
Food preparation	657	27.0	16.4
Cooks	2,073	24.0	17.4
Auto body repair	220	20.8	5.4
Health care services	2,680	11.5	29.4
Construction	6,253	17.4	7.0
Plasterers	58	46.9	12.2
Concrete work	117	38.4	13.9
Drywall	191	35.1	8.6
Tile	93	29.8	4.0
Painters	636	28.5	7.6
Brick and stone masons	235	24.5	11.2
Precision food manufacture	429	31.2	13.5
Meat cutters	223	38.1	13.5
Food batch makers	55	15.0	12.0
Manufacturing (textiles and general)	17,698	17.7	15.6
Sorter and grader	128	34.1	17.6
Packing	301	31.9	21.5
Cutting and slicing operators	138	24.3	16.3
Machine operator	2,515	20.4	16.6
Assemblers	1,135	18.6	15.5
Textile sewing operators	368	38.3	13.5
Textile other machine operators	197	26.1	14.7
Dressmakers	92	21.1	8.1
Upholsterers	62	28.2	10.9
Pressing	71	36.8	29.0
Dry cleaning	205	25.3	20.1

TABLE 2c BLS Labor Force Statistics of Selected Employed Persons by Detailed Occupation and by Race for 2001

Occupation Agricultural, Forestry, and Fishing	Total (thousands)	Hispanics Percent	Blacks Percent
Agriculture, forestry and fishing	3,245	21.5	5.0
Owners and managers	1,108	3.6	0.9
Agriculture and related activities	2.004	32.6	7.1
Other agricultural workers	734	42.8	3.6
Farm hands	671	42.3	3.7
Farm industrial work	1,269	26.7	9.1
Gardeners	876	29.5	11.1
Graders and sorters	51	68.0	7.9

SOURCE: Modified from BLS. Table 11. Employed Persons by Detailed Occupation, Sex, Race, and Hispanic Origin at<ftp://ftp.bls.gov/Pub/special.request/lf/aat11.txt>.

TABLE 3 Estimated Occupational Disease Morbidity for Blacks and Hispanics Working in the United States, 1992

Disease Categories	Estimated No. of Occupational Illnesses Attributed to Occupation (Includes all races and ethnicity)	Estimated No. of New Cases of Occupational Illnesses Attributed to Occupation for Blacks (11%)	Estimated No. of New Cases of Occupational Illnesses Attributed to Occupation for Hispanics (10%)
Cancer	66,790-111,130	7,347-12,244	6,679-11,113
Coronary heart disease	36,500-73,000	4,015-8,030	3,650-7,300
Cerebrovascular disease	5,050-14,400	556-1,584	505-1,440
Chronic obstructive pulmonary disease	150,000	16,500	15,000
Sub-Total	258,340-348,710	28,418-38,358	25,834-34,853
Unclassified Occupational Illness[a]	538,675	61,454	55,867
TOTAL	817,015 - 907,385	89,872 - 99,812	81,701 - 90,720

[a] This category includes BLS occupational illness for 1992 and occupational illness reported in the same year by the U.S. government employees (see Leigh et al., 1997).

SOURCE: Leigh et al. (1997). Percentages from BLS (1999).

particular, was likely not corrected sufficiently for known under-reporting in administrative record-keeping systems (Silverstein et al., 1997; Punnett, 1999). Thus, the totals more likely underestimate the true values.

For the purpose of estimating Latino work-related morbidity we assume that the percentage of disease in Latinos and African Americans correspond to their percentage in the civilian labor force (10 percent and 11percent respectively) (DOL, 1999). Applying the 11 percent estimate, the range of new cases for Latinos would be between 81,701 and 90,720 for the year 1992 (see Table 3). There is no reason to believe that this yearly estimate has changed since the reported numbers for 1992.

In addition, other occupational diseases such as occupational asthma (included under chronic obstructive respiratory disease in Table 3) are considered, the morbidity total grows substantially. Other researchers (Milton et al., 1998; Wagner and Wegman, 1999) have estimated that as high as 21 percent of all cases of asthma have an occupational etiology, yielding estimates above 100,000 prevalent cases of occupational asthma in the United States. This translates into at least 10,000 possible cases of occupational asthma among Latino workers in the U.S., if the 10 percent proportion for Latinos working in the United States.

A more focused look at the occupations with the greatest number of injuries illustrates the proportion of Latinos in those categories (see Table 4a). The U.S. Department of Labor, Bureau of Labor Statistics, identified the 10 job categories with the highest numbers of injuries and illnesses among 1,883,380 cases analyzed in 1997 (DOL, 1999). Latino workers were over-represented in at least three of the most hazardous job categories: janitors, laborers, and cooks.

TABLE 4a Percentage of Hispanics in 10 Occupations with the Larger Number of Occupational Injury and Illness 1997

	Number	
Occupation	Cases	Percent Hispanic
All occupations	1,833,400	10.2
Truck drivers	145,500	6.3
Laborers (Non-construction)	106,900	11.5
Nurses aides	91,300	7.6
Janitors and cleaners	45,800	20.0
Laborers (construction)	45,800	17.9
Assemblers	44,300	9.7
Carpenters	37,100	8.2
Cooks	31,500	12.8
Stock handlers	29,200	7.7
Welders and cutters	28,400	8.4

SOURCE: BLS (1999).

More specific data on rates of occupational injuries and illness appear in Table 4b. This table provides the incidence rate of injuries and illness per 10,000 full-time-equivalent workers. In general, the Standard Industrial Codes (SIC) that define the industry with higher incidence rates correspond to the SICs with the highest percentage of Hispanics. It is also remarkable that the greatest (> 70 percent) contributor to the incidence rate of occupational illness is repetitive trauma illness. The only exception is agriculture.

OCCUPATIONAL MORTALITY ESTIMATIONS

Fatal Occupational Injuries

The U.S. Department of Labor compiles every year the number of traumatic fatalities by race within the U.S. civilian workforce over 16 years old (DOL, 2000). Traumatic occupational fatalities are deaths that occur during employment or in the course of employment and are caused

TABLE 4b Incidence Rates per 10,000 Full-Time-Equivalent Workers of Non-Fatal Injuries and Illnesses by Selected Industries and Cases in 2000

Industry	SIC Code	Injury Rate[a]	Illness Rate[b]	Repetitive Trauma Illness Rate[b]	Percentage Hispanic in Industry[c]
All private		580	39.4	26.3	10.9
Agriculture	01-02	680	31.3	6.7	21.5
Metal mining	10	460	27.6	23	11.9
Coal mining	12	710	45.4	28.5	11.9
Construction	15	820	10.4	10.7	17.4
Non-ferrous foundries	336	1390	105.9	77.3	9.8
Auto stamping	3465	1260	267.9	240	9.5
Motor cars	3711	1440	831.2	727	17.7
Food products	20	1020	213	181	31.2
Meat products	201	1190	553.9	485	38.1
Meat packing	2011	1550	921	812	38.1[d]
Sausage and meats	2013	1160	311	274	31.2[d]
Poultry slaughtering	2015	990	433	378	31.2[d]
Men's trousers	2325	530	245	224	21.1

[a] The occupational injury incidence rates (IR) represent the number of injuries per 10,000 full-time workers and were calculated as:
IR = (N/EH) * 20,000,000
Where: N = number of injuries; EH = total hours worked by all employees during the calendar year and 20,000,000 = base for 10,000 full-time-equivalent workers (working 40 hours per week, 50 weeks per year) (Based on BLS Table1, OSTB12/18/2001, Incident Rates of Non Fatal Occupational Injuries by Industry Selected Cases at <www.bls.gov.iif/oshwc/osh/os/ostb1001.txt>).
[b] The occupational illness incidence rates (IR) are calculated as above. Data from: BLS Table S14. Nonfatal occupational illness incidence rates by industry and category of illness, 2000 at<http://www.bls.gov/iif/oshwc/osh/os/ostb1005.txt>.
[c] Data from USDOL #11. Employed persons by detailed occupation, sex, race, and Hispanic origin at <ftp://ftp.bls.gov/pub/special.requests/lf/aat11.txt>.
[d] Estimates based on (c) above.

by acute incidents related to the victim's occupation. The first three columns of Table 5 present these data for 1998. The proportion of fatalities occurring to Latinos (12 percent of 6,026) is greater than the expected proportion corresponding to the percentage of Latino workers in the working civilian population (10 percent of 132,684,000). If the fatality rate were proportionate to the number of Latinos in the workforce, 10 percent or 603, fatalities would be expected, while 700 were observed. The 97 deaths in excess over the expected number, or a relative risk (Gardner, 1989) of 1.18, indicate that Latinos are 18 percent more likely to die a traumatic death from injury on the job than whites and African Americans combined. A similar analysis could be made for the 2000 data. What appears more dramatic in the 2000 data is the substantial increase of mortality rate per 100,000 workers among Hispanics. Rates increased by 11.3 percent from 1999 to 2000. The proportion of Hispanics fatalities has also increased substantially (33 percent) when compared with the white rates. Fatality rates for blacks and whites have both decreased while the Hispanic rate increased.

TABLE 5 Occupational Traumatic Fatalities (Numbers and Rates) for White, Black, and Hispanic Working Populations for 1998[2] and 2000[1]

Race and Ethnicity	Fatal Injuries No. (percent)*		U.S. Civilian Workers over 16 years old (thousands) no. (percent)*		Rates per 100,000	
	1998	2000	1998	2000	1998	2000
White	5,016 (83)	4,240 (72)	111,683 (84)	113,475 (81)	4.491	3.737
Hispanic[3]	700 (12)	815 (14)	13,381 (10)	14,492 (10)	5.231	5.624
Black	591 (10)	574 (10)	14,795 (11)	15, 334 (11)	3.995	3.743
Total	6,026	5,915	132,684	140,863	4.541	4.199

NOTES:
1. The Hispanic fatality rate grew by 11.3 percent from 1999 to 2000 (5.213 to 5.623). In addition, in 1999 the Hispanic fatality rate was 16.7 percent greater than the white fatality rate (5.231 compared with 4.484). The disparity between the Hispanic and white fatality rates grew to 33.0 percent in the year 2001 (5.623 compared with 3.763).2. Percentages are greater than 100 because Hispanic workers can also be identified as black and white.

SOURCES: BLS (1999, 2001).

Because, since the Latino classification in the Bureau of Labor Statistics includes members of every race (and thus the percentages sum to more than 100 percent), no valid statistical test could be applied to measure differences in fatal occupational injury rates by race/ethnicity. To bring some context to these traumatic fatalities, the Bureau of Labor Statistics reports the event or exposure to which each death is directly attributed (i.e., the underlying cause) (see Table 6).

Transportation incidents are by far the principal cause of traumatic occupational fatalities (44 percent), followed by homicide (16 percent) and being struck by objects and equipment in work settings (15 percent)(DOL, 2000). Unfortunately the BLS statistics only provide the distribution of all deaths by race/ethnicity, without stratification by event or exposure. In order to have a benchmark the number of traumatic deaths were estimated within each race/ethnicity by event or exposure, assuming the same percentages as the total number of deaths. These numbers are thus only an approximation, since the distribution of occupational titles differs between African Americans, whites and Latinos.

In recent studies estimates of fatal injury lifetime risk accumulated over the life of the worker have been calculated (Myers et al., 1998). In 1998 the NIOSH published a study comparing experiences of the highest working lifetime risks by race/ethnicity that permits a direct comparison between Latinos, African Americans and whites (see Table 7).

All comparisons were made between lifetime risks of African Americans and whites in the same occupation/industry categories. In Table 7 the descending list of the highest fatal working lifetime risk for Latinos in five occupations within five industries is compared with the same occupation and industry lifetime risk experiences of African Americans and whites. Latinos had the highest lifetime risk of fatalities by homicide among cashiers in gas stations, guards in security services, and collisions as truck drivers. Latinos also had the second highest lifetime risk of homicide as cab drivers and employees of grocery stores (Myers et al., 1998).

TABLE 6 Occupational Traumatic Fatalities by Event or Exposure, 1998: Hispanic (Latino), Black and White Populations

	Estimation of Fatalities by Race and Ethnicity		
Event or Exposure	BLS 1998	Percent Events	Hispanic
Transportation incidents	2,630	44	308
Assault, violence	960	16	112
Contact objects and equipment	949	15	105
Falls	702	12	84
Exposure electricity toxins	572	9	63
Other	213	3	28
Totals (from Table 4)	6,026		700

NOTES:
1. BLS obtained the distribution of "Percent Events" over all races and ethnicities. BLS reported the distribution of race and ethnicity of all the 6,026 fatalities as "Percent Events" with no breakdown by race and ethnicity.
2. The estimation of numbers of deaths by event and by race and ethnicity was based on "Percent Events" of all fatalities (i.e., calculation of estimate of transportation fatalities of Hispanics: 700 * 0.44 = 308).
SOURCE: BLS (1999).

TABLE 7 Highest Working Lifetime Fatality Risk for Homicides and Transportation Incidents for Hispanics, Black and White Workers – (Accumulated Data from 1992-96 for Deaths Greater than Five)

			Hispanic		Black	
Industry	Occupation	Event	# Deaths	Lifetime Risk	# Deaths	Lifetime Risk
Cab or transportation	Driver	Homicide or shooting	46	49.5	137	66.7
Gas station	Cashier	Homicide or shooting	12	13.1	6	4.5
Grocery store	Employee	Homicide or shooting	52	12.2	30	7.2
Security service	Guard	Homicide or shooting	31	9.5	41	4.5
Truck or transportation	Driver	Collision	28	4.9	42	4.2

NOTES:
1. Working lifetime risk (WLTR) in units of deaths per 1,000; 45-year working lifetimes
$WLTR = [1 - (1 - R)^y] \times 1000$
where R = Ratio of the average annual number of work-related fatal injuries among workers in a given group to average annual employment in that group; y = Years of exposure to work-related fatal injury risk.
2. Working lifetime is assumed to start at 20 and end at 65 years.
3. NA = not applicable.

SOURCE: Modified from Myers et al. (1998).

Fatal Occupational Diseases

Fatal occupational disease is defined as death due to a disease that is either caused or exacerbated by substances, physical conditions, or other hazardous exposures on the job. The best available estimates of fatal occupational disease in the U.S. general population were published recently by Leigh et al., (1997). Their estimates considered six groups of occupational diseases: cancer and cardiovascular, renal, neurological, pneumoconioses, and chronic respiratory diseases in which occupational exposures were the prime contributor to death (see Table 8). They estimated that 46,800 to 73,600 deaths per year in the United States due to occupational diseases (see Table 9).

TABLE 8 Estimated Fatal Work-Related Disease for the White, Black, and Hispanic Working Populations

Race/Ethnicity	Fatal Work-Related Disease Range	Percent
White	38,844-73,600	83
Black	4,680-7,360	10
Hispanic (Latino)	5,148-8,096	12
Total	46,800-73,600	100

NOTE: Percentages are greater than 100 because overlap of counting racial classifications (e.g., Hispanics can be double counted as whites and blacks.

SOURCES: BLS (1999) Percentages of traumatic fatalities as reported by BLS for 1998. Modified from Leigh et al. (1997).

TABLE 9 Estimated Fatal Work-Related Illness Distributed by Disease

Causes of Death	Estimated No. of Deaths Attributed to Occupation (all races and ethnicity)	No. of Deaths of Hispanics Attributed to Occupation (12%)	No. of Deaths of Blacks Attributed to Occupation (10%)
Cancer	31,025-51,706	3,723-6,204	3103-5171
Cardiovascular and cerebrovascular disease	5,092-10,185	611-1,222	509-1,019
Chronic respiratory diseases	9,154	1,099	915
Pneumoconioses	1,136	136	114
Nervous system disorders	269-806	32-96	27-89
Renal disorders	223-689	27-83	22-76
Total	46,800-73,600	5,520-8,832	4,680-7,360

SOURCES: Leigh et al.(1997); BLS (1999).

These estimates are more conservative than the previously published estimates of 100,000 deaths annually (Ashford, 1976). Using the BLS estimate for 1999 (DOL, 2000) stating that 12 percent of the traumatic fatalities in the U.S. civilian workforce were of Latino workers (see Table 5), we can roughly estimate that 5,520 to 8,832 Latinos die from occupational diseases annually. The same 1997 estimates by Leigh et al. of deaths by occupational disease are distributed by cause of death using the percentages of death for traumatic fatalities (12 percent for Latinos, 10 percent for African Americans). The results appear in Table 8.

QUALITY OF DATA ON LATINO WORKERS

All federally generated data on morbidity and mortality are based on reported counts for a limited number of workers in the private sector only (excluding self-employed workers). For any overall estimation, adjustments (e.g., additional data sources) need to be made to arrive at total counts for the entire work force. Unreported and uncounted cases generally lead to underestimation of occupational morbidity and mortality. A second estimation is required because of the inadequate recording of race/ethnicity, necessitating the assumption that the risk of occupational disease and injury is the same in the U.S. Latino workforce as in the general U.S. workforce. Because evidence strongly suggests that Latinos are over-represented in the more hazardous jobs and are at higher risk, the true numbers are likely to be even higher.

Landrigan and Markowitz (1989) estimated the degree of under-reporting by comparing independently generated estimates of occupational disease mortality and incidence with the actual numbers of cases reported by the Workers' Compensation Board. In New York State, only 3 percent to – 5 percent of the estimated number of deaths due to occupational diseases was reported by the Workers' Compensation Board as being of occupational etiology. Similarly, 11 percent to – 38 percent of the estimated number of incident cases of occupational disease were reported by the Workers' Compensation Board as occupational. Many of the estimated 95 percent to– 97 percent of occupational disease deaths and 62 percent to – 89 percent of incident cases of occupational disease that were not reported as occupational probably represent unrecognized epidemics.

COSTS OF OCCUPATIONAL MORBIDITY AND MORTALITY

Missed and delayed diagnoses of occupational disease produce substantial and avoidable costs in the form of time lost from work, decreased productivity, economic hardship and unemployment, increased burden on an already overloaded health care system, inappropriate diagnostic testing or treatment, prolonged duration of disease, progression of reversible disease to chronic irreversible disease, and most important, suffering, disability, and death. The hidden costs are borne mainly by workers and their families; employers; union benefit funds; city, state, and federal governments; and in some cases, medical insurance companies (who pass the costs on to their other clients). These costs contribute to the rising costs of medical insurance. In addition, the missed diagnoses are not reported as occupational diseases and are thus invisible to the occupational health surveillance systems, delaying recognition and resolution of the problem of work-related disease.

Leigh et al. (1997) estimated that the total cost of occupational morbidity and mortality in the United States reaches $171 billion per year: $65 billion in direct costs plus $106 billion in indirect costs. The grim share of this cost from the Latino working population is approximately 12 percent (Leigh et al., 1997) of $171 billion, or $21 billion dollars each year.

WORK-RELATED DISEASES IN LATINO WORKERS

Which work-related diseases occur in Latino workers in urban areas of the United States? Specific statistics of incidence, prevalence, and mortality data for Latino work-related diseases have not been compiled. However, some examples of work-related diseases that occur in industries and occupations that employ many Latino workers are listed as "occupational sentinel health events" (Rutstein et al., 1983; Mullan and Murphy, 1991). Occupational sentinel health events are diseases or causes of disability or death (with ICD-9 codes) that satisfy defined criteria for literature-supported associations with specific toxic substances, industries, or occupations. In individual cases a detailed occupational health clinical evaluation (as discussed in several occupational medicine textbooks (LaDou, 1998; Rosenstock, 1996; Levy, 1999; Rom, 1998) is indicated once a work-related disease is suspected, in order to clinically confirm or rule out work-relatedness. In addition, sixteen reviews of clinical evaluation for specific occupational diseases have recently been published in the American Journal of Industrial Medicine (January 2000).

LATINO WORKERS RISK OF OCCUPATIONAL DISEASE

Table 4a shows the blue-collar occupations with the highest number of occupational illness and injury nationwide. These 1997 data show how the percentage of injuries/illness exceeded the representation of Latino workers in three categories (janitors, laborers, and cooks) (DOL, 1999). Another approach to judging the magnitude of the problem of occupational disease in Latino workers is to examine data on the rates of occupational disease in various industries and the presence of Latinos in these. A sample of that strategy appears on Table 4a.

SWEATSHOPS IN THE GARMENT INDUSTRY

Sweatshops have been discussed in two documents published recently by the U.S. General Accounting Office, which defines them as "businesses that regularly violate both safety or health and wage or child labor laws" (GAO, 1988); sweatshops and are also defined more loosely as "chronic labor law violators." Construction firms, farms, and homework (e.g., piecework apparel or electronics manufacturing in the home) are included in the definition.

By definition sweatshops are hazardous workplaces. The occupational hazards encountered in sweatshop work in the garment industry include ergonomic hazards (e.g., repetitive motions, awkward working postures, vibrating tools such as fabric cutters, falls from ladders), airborne hazards (e.g., high concentrations of dusts, poorly ventilated dry-cleaning solvents and fumes from glues and fabric treatments like formaldehyde), temperature extremes, and skin contact with irritant and allergenic substances. Occupational diseases and injuries prevalent among apparel sweatshop workers include musculoskeletal or cumulative trauma disorders like back, neck and shoulder pain and carpal tunnel syndrome (Punnett et al., 1985; Sokas et al., 1989), contact dermatitis, occupational asthma and bronchitis, vibration-induced Raynaud's phenomenon, and acute and chronic toxicity from solvents and other toxic chemicals.

A poll of 53 federal enforcement officials, published by the U.S. General Accounting Office (1988), showed that sweatshops were reported in significant numbers in 47 of the 50 states, most commonly in the apparel, restaurant, and meat processing industries. Major concentrations were found in large cities, with New York the most intensively studied. At the national level Latinos were thought to make up the majority of sweatshop workers in both restaurant and apparel industries, followed by Asians and African Americans.

No national estimates of the number of sweatshop workers have been published, but local estimates from several sources have been reported. The director of the N.Y. State Department of Labor Apparel Industry Task Force estimated that 4,500 of the 7,000 apparel factories and shops in New York City were sweatshops, and that over 50,000 workers were employed in this sector. Other sources estimated the number of apparel sweatshops at 3,000, also with over 50,000 sweatshop workers (GAO, 1988). In New Orleans an estimated 25 percent of the 100 apparel firms (employing 5,000 or more workers) were multiple labor law violators. The only available estimate for restaurant workers comes from an official in Chicago, who estimated that half of their 5000 restaurants (employing 25,000 or more workers) were chronic labor law violators (GAO, 1988).

Sweatshops have proliferated because of continuing social and economic factors. The reasons for the existence of sweatshops cited by more than 50 percent of the federal officials (GAO, 1988) were (in decreasing order of response):

1. available supply of an immigrant workforce
2. labor-intensive nature of the industries
3. low profit margin in these industries
4. too few inspectors
5. weak or nonexistent unions
6. inadequate penalties for infractions

Not mentioned but probably important is the fact that these industries require skills that are either already possessed by a large proportion of the immigrant workforce or can be learned on the job with little formal training. This does not imply that these are low-skilled jobs; Fernandez Kelly gave a graphic description of the high level of skill and productivity needed even to approach earning the minimum wage in a piecework payment system (Fernandez-Kelly, 1983). Sweatshops have also been identified as a system of production that invariably inflicts violence on minorities, either traumatic violent deaths and injuries (fires, explosions, accidents) or chronic occupational diseases that although extended in time are not less violent (Moure-Eraso, 1999).

MAQUILADORA INDUSTRY AS MODERN SWEATSHOPS

Closely related to sweatshops are maquiladoras, manufacturing and assembly factories along the Mexican side of the U.S.-Mexico border. The maquiladora produces a variety of goods (from Zenith television sets to computer boards and wire harnesses) earmarked exclusively for export to the United States. More than one million workers work in this industry (Cedillo, 1999). These workers are exclusively Mexican citizens working in the Mexico side of the border. The community and occupational health impacts have been studied only in the last 10 years of its explosive growth. A community study of 267 maquiladora workers, mostly female, in the cities of Matamoros and Reynosa found evidence of musculo-skeletal disorders related to working conditions (pace of work, poor workplace design, and ergonomic hazards). Acute health effects were also identified compatible with chemical exposures. Although other chronic diseases were not apparent, the high presence of musculo-skeletal disorders was a striking result for a very young workforce (average age 25 years) (Moure-Eraso et al., 1994, 1997).

BARRIERS TO PREVENTION OF OCCUPATIONAL DISEASES

The evidence presented above strongly suggests that occupational disease in the Latino workforce is a common, severe, and preventable problem. In order to suggest rational solutions to the problem it would be helpful to explore some of the barriers to prevention of occupational diseases as well as some of the reasons that Latino workers tend to be over-represented in hazardous jobs.

In principle all occupational diseases are preventable. Prevention depends on identifying the causal exposure(s) and eliminating or reducing these exposures until no more workers get sick. Effective prevention requires timely identification and elimination or control of the causative exposures. It is preferable to try to prevent the disease from occurring at all, rather than allowing it to occur and then trying to rehabilitate the person and provide monetary compensation for damage already done.

Several types of barriers currently exist to prevent implementation of effective prevention strategies. A lack of scientific understanding of the causal agents and their mechanisms is one important barrier, and research must be ongoing to identify hazardous agents and work situations. For example, only about 10,000 of the 60,000 commercially used chemicals have been tested for toxicity in animals (LaDou, 1998), and very few of these have been studied epidemiologically in humans. Understanding of the cellular and molecular mechanisms of toxicity is clearly an important component in developing new approaches to treatment and prevention of occupational disease. Even so, results of epidemiological studies can often be used to prevent occupational disease even without detailed knowledge of mechanism. An example is the identification of asbestos-related diseases using epidemiological methods. This knowledge led to the reduction in asbestos-related disease through regulation and decreased use of asbestos in the United States. Other situations may not be so clear-cut, and it may require more sophisticated epidemiological techniques (e.g., preventive intervention trials) or basic understanding of the toxicology to identify the causative agent.

Although an incomplete understanding of the biomedical and epidemiological etiologic agents causing specific occupational diseases are important factors, the main barriers to prevention have been economic, political, legislative, and social. Frequently workers cannot afford to turn down a job for which they are qualified, simply because it is hazardous; they often are unable to leave a job they know is hazardous, because they do not have the financial resources to stop working for the time required to find another job, retrain for another job, or to apply for and receive Workers' Compensation payments. It is an unfortunate fact that Workers' Compensation payments for occupational diseases generally take more than six months, and often over a year, to begin, even in the most clearly documented cases. Although this barrier affects relatively affluent workers as well, it affects workers with less financial resources more severely. In addition, the threat of prolonged unemployment weighs heavily on members of communities where jobs are scarce. The need to continue supporting a family frequently serves as an irresistible pressure on workers to endure concrete physical discomforts or more abstract elevated risks of occupational diseases. As a result workers in these circumstances may present with advanced or late-stage occupational diseases and may be resistant to quitting their jobs even when strongly counseled to do so by a physician. This economic and bureaucratic trap is a major barrier to preventing occupational illness in Latino workers.

Employers who are ethical, well informed about occupational health, and sensitive to the particular needs of their workforces can play a key role in providing safe and healthy workplaces. Labor—anagement health and safety committees can be helpful in raising issues and resolving them before serious health effects occur. Employers sometimes believe that improvement of working conditions will be prohibitively expensive and will not make the investment, even if their workers are getting sick from exposures on the job. In spite of this common perception that safe and healthy working conditions cost too much, sometimes the cost of directly eliminating the hazard is less than the combined long-term costs of decreased productivity due to time lost from work and low morale, increased Workers' Compensation premiums, and fines from enforcing agencies like the OSHA or the EPA. Sometimes a "win-win" solution can be found that improves the working conditions without incurring unmanageable costs and may even increase productivity (Friedman-Jimenez and Claudio, 1998).

Lack of access to comprehensive clinical occupational health services is another barrier to recognition and prevention of occupational diseases. Access is difficult for most workers, particularly so for Latino workers. Evidence for this is indirect, since direct measures of access to these services have not been published. The increased risk of occupational disease combined with poor access to clinical, preventive, and educational occupational health services suggests that the public health impact of interventions to correct this situation would be particularly great.

Overcoming the socioeconomic and political barriers to prevention may prove to be an even greater challenge for Latino workers than for white workers, and will not occur until the problem is recognized and adequate resources are committed to a rational solution.

A COMPREHENSIVE APPROACH TO ADDRESSING OCCUPATIONAL DISEASE IN LATINO WORKERS

The evidence strongly suggests that preventable occupational diseases cause substantial mortality and morbidity in the U.S. Latino population, and that this problem is not being recognized or addressed adequately. Admittedly the evidence is fragmented and of variable quality, but it is more than sufficient to drive us to begin addressing this problem now. The economic and non-economic costs of not addressing the problem are great. The OSHA has not been effective in identifying hazardous exposure situations or epidemics of occupational disease, as we have seen in the report on sweatshop inspections. We need to address the problem directly and break the cycle of "no services → no data → no services", by making comprehensive occupational health services more accessible to Latino workers and simultaneously documenting the epidemics with careful clinical, epidemiological, and surveillance studies of working populations that include significant numbers of Latino workers. To facilitate this process recommendations are offered for a comprehensive approach to addressing this problem. These recommendations fall into six categories:

1. primary prevention intervention approaches;
2. clinical services;
3. educational approaches;
4. research and surveillance;
5. unionization and organization of workers; and
6. legislation and regulation of hazardous workplace exposures.

To have a significant impact on the occupational health of Latino workers efforts must be made in all six areas. Although occupational diseases may disproportionately affect Latino workers, the aim of preventive programs should be to reduce hazards for all workers, not simply redistribute the hazards more "equitably." These recommendations are intended to supplement broader, ongoing efforts to improve health and safety conditions in the workplace.

Primary Prevention Interventions

Occupational health is looking more and more to primary prevention interventions as the way to evolve from a paradigm of control of occupational exposures to a new one of prevention (Moure-Eraso, 1999; Quinn et al., 1998). For example, traditional industrial hygiene has at the top of the hierarchy of engineering interventions the fabrication of local exhaust ventilation systems to "control" chemical exposures in work environments. Such strategies are proving to be very problematic in the long run. The ventilated (exhausted) toxic substance extracted from the worksite becomes an environmental contaminant in the community. This method of control generates toxic pollutants (gases or tiny particles) that need to be treated as either a hazardous waste or an air pollutant. In environmental science that approach is defined as an end of pipe solution equivalent to waste disposal. New strategies for engineering interventions are being developed that avoid this type of risk transfer. The most obvious one for primary prevention purposes is raw material substitution. If this is not viable, toxic use reduction or source reduction by process changes are the strategies of choice (Moure-Eraso, 1999; Ellenbecker, 1996).

Housing in minority communities often clusters at factories and emitters of pollutants. Risk shifting from the point of production can be avoided or at least mitigated by such primary prevention strategies as substitution, source reduction, closed-loop recycling, improvement of maintenance, process modernization, reformulation of products, and improvements of housekeeping and training changes (Moure-Eraso, 1999; Ellenbecker, 1996). All these interventions need to occur at the point of production and with full participation of the parties involved (i.e., workers, companies, and occupational health professionals).

CLINICAL OCCUPATIONAL HEALTH SERVICES

Access to comprehensive clinical occupational health services is probably the major determinant of the success of medical treatment for occupational diseases, as well as of surveillance and secondary prevention programs. In New York City and other urban areas in the United States, working Latinos have even less access to these services than the general working population. Access is determined mainly by the existence of local occupational health clinics and primary care providers knowledgeable in occupational medicine, as well as existence of other referral sources, such as concerned unions, and workers and businesses aware of occupational health. Occupational diseases are usually only recognized when the diagnoses are specifically considered by clinicians with some training in occupational medicine, or when suggested by an educated worker or patient.

Comprehensive occupational health services should include:

1. diagnosis and symptomatic treatment of the medical condition;
2. determination of whether the medical condition is work-related and identification the causative exposure(s) as specifically as possible;
3. evaluation of workplace conditions, including inspection of the workplace if necessary and feasible;
4. the capability to mount a group medical screening of co-workers from the same workplace with similar exposures, if indicated;
5. education of the worker or patient as well as the employer and union regarding occupational hazards;
6. gaining the cooperation of the employer and the union in addressing health and safety issues on the job;
7. removal or control of the hazardous exposure by materials substitution, engineering controls, personal protective equipment. or if this proves impossible, removal of the worker from the workplace;
8. filing Workers' Compensation applications when appropriate;
9. if necessary and with the worker's informed consent, reporting hazardous workplaces to the OSHA, NIOSH, EPA, or the appropriate regulatory or research agency;
10. reporting all cases to the appropriate surveillance program if one exists (e.g., Occupational Disease Registry of the New York State Department of Health, or the NIOSH Sentinel Event Notification for Occupational Risk (SENSOR) program (Baker, 1989);
11. facilitating vocational rehabilitation and job retraining for workers disabled by occupational diseases (e.g., workers with allergic sensitizations to substances present at the worksite);
12. educating primary care providers about recognition, basic management, and referral of patients with likely occupational diseases.

Clinical services for most Latinos begin with primary care providers in the Latino communities, including practitioners, health centers and providers in hospital outpatient departments and emergency rooms. The majority of primary care providers have never been trained to recognize occupational diseases and do not know how or where to refer a patient they think may have an occupational disease. Education of hospital-based providers is important, but education of the community-based providers is crucial, since they are often the first or only accessible source of health care and advice. If the first contact providers do not recognize the occupational etiology of the illness, it is unlikely that the correct diagnosis will be made at all. These providers should have a basic knowledge of the diagnoses they might encounter that are likely to be work-related, the occupational sentinel health events discussed earlier. Appropriate referral channels must be accessible to the providers to follow up on possible work-related diagnoses. The Association of Occupational and Environmental Clinics (AOEC), based in Washington, D.C., has over 60 clinics across the United States available for referrals from primary care providers. The New York State Department of Health supports a network of eight occupational health clinics that see patients referred by community health care providers, employers, unions, and other sources, including self-referrals.

EDUCATIONAL APPROACHES

Much is already known about workplace hazards and prevention (see Rom, 1998; Rosenstock and Cullen, 1996; La Dou, 1998; Levy and Wegman, 1999; DiNardi, 1998). Application of this knowledge to improve public health requires effective education of professionals, workers, and employees.

Comprehensive, culturally and linguistically competent occupational health education programs should be accessible to all workers in specific industries and occupations. To be accessible to Latino workers these programs need to be in Spanish and English and should include Right-to-Know education, other health and safety training, and Spanish translations of relevant material safety data sheets (MSDSs). An excellent book by A. Kimball (1990) which lists and reviews 289 Spanish-language occupational health and safety materials for workers, is a useful aid in planning and conducting worker education programs for Spanish-speaking Latino workers. Literature on empowerment approaches to worker health and safety education by Wallerstein et al. (1992 and 1993) emphasizes active involvement of workers in creating solutions to health problems in their own workplaces.

Employers are often not well versed in health and safety issues, even related to hazardous substances or conditions in their own facilities. Educational programs to make them aware of appropriate health and safety practices and to sensitize them to worker concerns and perspectives will facilitate improvements in health and safety conditions. Many corporate and union health and safety departments have been downsized or eliminated. Occupational health programs in academic centers or in the community (e.g., Committees for Occupational Safety and Health) could play increasingly important roles in meeting occupational health training needs.

Limitation of job opportunities due to lack of education is important as an independent factor, and in combination with other factors, in increasing the risk of occupational diseases in Latino workers. Improvement of both quality of education and numbers of high school and college graduates in Latino communities would help lower the risk of occupational disease by opening up opportunities for less hazardous jobs and by empowering workers to improve their working conditions.

RESEARCH AND SURVEILLANCE

Sufficient evidence exists to justify addressing the problem of occupational disease in Latino workers immediately. Nevertheless, for this issue to compete successfully with other high-priority issues for funding and research and clinical talent, objectives of this research would be (1) to determine which occupational health issues are most pressing; (2) to document the extent and severity of the problem; and (3) to develop effective and practical solutions.

The NIOSH has called to the occupational health community to develop a National Occupational Research Agenda (NORA) to address the financial and human challenges brought about by the large burden of work-related disease, injury, and death. This agenda has developed 21 research priorities that includes categories of special importance to the Latino community; special populations at risk; occupational health services; social and economic consequences of workplace illness and injuries; and innovative methods of hazards and health surveillance (NIOSH, 1998). NIOSH has organized periodic meetings with more than 500 organizations and individuals to obtain input for the development and guidance of research programs under NORA.

Improving access of Latino workers to occupational medicine clinical services would help greatly in documenting the prevalence and incidence of currently undiagnosed occupational diseases in Latinos. A successful statewide network of occupational health clinics coordinated by and reporting to the state department of health and supported by funds from Workers' Compensation was started in New York State in 1989.

State and federal occupational disease surveillance programs should include standardized ethnic classifications and should provide data on Latino workers. In 1984 the collection of occupational disease data in the United States was described as "fragmented, unreliable and 70 years behind communicable disease surveillance" (Committee on Government Operations, 1984). Both occupational disease surveillance and communicable disease surveillance have advanced since 1984 but a large gap remains between data collection in these two areas.

Principles of effective occupational health surveillance have been well summarized by Markowitz in Rom (1998), who wrote that "occupational health surveillance entails systematic monitoring of health events and exposures in working populations in order to prevent and control occupational hazards and associated diseases and injuries." He went on to list four essential components of occupational health surveillance summarized by Markowitz in Rom 1998, who:

1. gather information on cases of occupational diseases and injuries and on workplace exposures.
2. distill and analyze the data.
3. disseminate organized data to necessary parties, including workers, unions, employers, government agencies, and the public.
4. intervene on the basis of data to alter the factors that produced these health events and hazards. (See "Primary Prevention Interventions" above).

Consensus among authorities on occupational health surveillance holds that intervention must be an integral part of any surveillance program, in addition to the first three components mentioned above. It is clear from these principles that accessible occupational health clinical and educational services are necessary and prerequisite to establishment of an effective program of occupational health surveillance.

Unionization, Worker Organization, Legislation, and Regulation of Workplace

Ideally employers would provide healthy and safe working conditions without intervention from the outside. Some companies have identified worker health and safety as a priority and have in-house occupational medicine departments or have hired consultants to provide occupational health services. However, it is fairly common for employers not to provide healthy and safe working conditions. In these situations a union can facilitate the process of improving the health and safety conditions in the workplace.

Organized workers are better able to avoid or reduce toxic and dangerous exposures in the workplace, with less vulnerability to being fired by an unscrupulous employer. Health and safety laws do not prevent an employer from firing a worker for becoming sick from their job, filing for Workers' Compensation, requesting appropriate protective equipment, or for simply complaining about dangerous working conditions. The protections under the law that do exist are frequently circumvented when the employer asserts that the worker was terminated for economic or other reasons not related to health, and the burden of proof rests on the worker to demonstrate otherwise.

Like other workers in the United States, only about one in six of Latino workers is a member of a union. Organizing Latino workers can be a crucial step in improving compliance with OSHA guidelines and bringing about safe and healthy working conditions, as we saw in the coated-fabric factory example. A strong union health and safety committee can often be effective in facilitating this process, especially when management cooperates and forms a labor and management health and safety committee.

CONCLUSIONS

The available evidence is inadequate to quantify the prevalence and incidence of occupational diseases and injuries in Latino working populations (although more and more sound estimates are becoming available) (Leigh et al., 1997) at national and state levels. The current estimates suggest that the overall risk of occupational diseases and injuries among Latino workers is beyond what should be expected. In spite of obvious inadequacies available evidence is certainly sufficient to justify primary prevention interventions in the occupations for which we have identified morbidity and mortality excesses. It also warrants a greatly accelerated investigation of the problem, specifically as it pertains to Latino workers and high-risk industries that employ them.

Local epidemics of occupational diseases among Latino workers in particular industries have occurred and undoubtedly continue to occur. These epidemics are sometimes severe and could be readily observed if we looked for them.

It is clear that primary prevention interventions are needed in the workplace, such as substitution of problem chemicals; process changes to address the sources of chemical and physical hazards; and other engineering interventions. In addition, secondary prevention interventions must also be conducted, including a rapid and substantial increase in clinical, preventive, and educational occupational health services accessible to Latino workers. Because most occupational diseases remain undiagnosed and unreported by the current medical system, improving the recognition, diagnosis, and reporting of occupational diseases by primary care, subspecialist, and occupational medicine providers will be critical. Recognition and clinical diagnosis of occupational diseases are prerequisite to adequate reporting, so improving access to clinical occupational health services will be a necessary step in this process.

Although the focus of this chapter has been Latino workers, it is clear that in order to adequately address the problems of occupational disease and injury among Latino workers, any solution must address these problems among high-risk workers of all races and ethnicity. Addressing local epidemics haphazardly, if they happen to be discovered, will never be an adequate solution to this problem. A more humane, scientifically valid, and ultimately, cost-effective strategy for addressing this problem would be an integrated program of occupational health surveillance, accessible clinical occupational health services, careful epidemiological research, worker and health care provider education, collaboration with cooperative employers, unionization of Latino workers, legislative reform, and improved enforcement of regulatory standards.

REFERENCES

Ashford N, Crisis in the workplace: Occupational Disease and Injury. MIT Press, Cambridge MA. 1976.

Baker E. Sentinel Event Notification System for Occupational Risks (SENSOR): The concept. Am J Pub Hlth, 79(Supplement): 18-20, 1989.

BLS (U.S. Bureau of Labor Statistics). National Census of Fatal Occupational Injuries, 1998. Washington DC. 2000.

BLS Report on the American Workforce 2001. USDOL, Washington DC, 2001.

BLS Table 1, OSTB12/18/2001, Incident Rates of Non Fatal Occupational Injuries by Industry Selected Cases. www.bls.gov.iif/oshwc/osh/os/ostb1001.txt).

BLS Table S14. Nonfatal occupational illness incidence rates by industry and category of illness, 2000. http://www.bls.gov/iif/oshwc/osh/os/ostb1005.txt.

BLS Occupational Injuries and Illness.1997. Washington, D.C., 1999.

Cedillo L, Psychosocial Risk Factors Among Women Workers in the Maquiladora Industry in Mexico. Doctor of Science Dissertation, University of Massachusetts Lowell, Lowell. 1999.

Census Bureau (NP/TS/A). Projections of the resident population by race. Washington DC. 2000.

Committee on Government Operations. Occupational illness data collection: Fragmented, unreliable and seventy years behind communicable disease surveillance. House Report 98-1144, Oct. 5, 1984.

DiNardi S (Ed). The Industrial Environment—Its Evaluation and Control. American Industrial Hygiene Association (AIHA). Fairfax, VA. 1998.

DOL (Department of Labor). Standard Industrial Classification Manual, Washington DC. 1987.

Ellenbecker MJ. Engineering controls as an intervention to reduce workers exposure. Am J Ind Med. 29; 303-07. 1996.

EPA (Environmental Protection Agency). Sociodemographic Data Used for Identifying Potentially Highly Exposed Populations. EPA/600/R-99/060, July 1999.

Fernandez-Kelly, MP. Maquiladoras: The view from the inside. Ch. 14, Women and Industry in Mexico's' Frontier. State University of New York, Albany. 1983.

Friedman-Jiménez, G., Claudio L. Environmental Justice. In Rom WN (ed): Environmental and Occupational Medicine. Third Edition. Philadelphia. L-R. 1998

Frumkin H, Walker ED, Friedman-Jiménez, G. Minority Workers and Communities. In: Special Populations. State of the Art Reviews, Occupational Medicine. Volume 14, No.3. Hanley&Belfus, Philadelphia.1999.

GAO (General Accounting Office) briefing report to the Honorable Charles E. Schumer, House of Representatives. "Sweatshops" in the U.S.: Opinions on their extent and possible enforcement options. GAO/HRD-88-130BR, 1988.

GAO briefing report to the Honorable Charles E. Schumer, House of Representatives. "Sweatshops" in New York City: A local example of a nationwide problem. GAO/HRD-89-101BR, June, 1989.

Herbert R, and Landrigan PJ. Work Related Death: A continuing Epidemic. Am J public Health, 90:541-45. 2000.

Herbert R, Luo J, Marcus M, Landrigan PJ, Plattus B, O'Brien S. The failure of workers' compensation statistics to monitor work-related illness in garment workers. APHA abstract, presented at APHA Annual Meeting, 1989.

Kimball, A. A Workers' Sourcebook: Spanish language health and safety materials for workers / La Fuente Obrera: Materiales de salud y seguridad en español para trabajadores. Labor Occupational Safety and Health Program, Center for Labor Research and Education, Institute of Industrial Relations, UCLA. 1990.

LaDou, J. Occupational Medicine. Appleton and Lange. 1998. New York, NY.

Landrigan, P.J., Markowitz, S. Current magnitude of occupational disease in the United States: Estimates from New York State. Ann. NY Acad. Sci 572:27-45; 1989.

Leigh JP, Markowitz SB, Fahs M, Chonggan Shin, Landrigan PJ. Occupational Injury and Illness in the United States: Estimates of Costs, Morbidity and Mortality. Archives of Internal Medicine; 157: 1557-68. 1997.

Levy, B.S., Wegman, D.H. (eds). Occupational Health: Recognizing and preventing work-related disease. 3rd edition. Boston: Lippincott, Williams & Wilkins, New York, NY. 1999.

Milton DK, Solomon GM, Rosiello RA Herrick RF. Risk and Incidence of asthma attributable to occupational exposure among HMO members. Am J Ind Med. 33. 1: 1-10. 1998.

Moure-Eraso R. Sweatshops. In: Violence in America: An Encyclopedia. Gottesman R (ed). Vol 2: 243-4. Charles Scribber, New York. 1999.

Moure-Eraso R, Wilcox M, Punnet L, MacDonald C, Levenstein C. Back to the Future: Sweatshop Conditions on the Mexico-U.S. Border. I. Community Health Impact of Maquiladora Industrial Activity. Am J Ind Med. 25: 311-20. 1994.

Moure-Eraso R, Wilcox M, Punnet L, MacDonald C, Levenstein C. Back to the Future: Sweatshop Conditions on the Mexico-U.S. Border. I I. Occupational Health Impact of Maquiladora Industrial Activity. Am J Ind Med. 31: 587-99. 1997.

Moure-Eraso, R. Avoiding the transfer of risk: Pollution prevention and occupational health. In: Levy, B.S., Wegman, D.H. (eds). Occupational Health: Recognizing and preventing work-related disease. 3rd edition. Boston: LW&W, New York. 1999.

Mullan RJ, Murthy LI. Occupational Sentinel Health Events: An up-dated list for physician recognition and public health surveillance. Am J Indust Med 19:775-799. 1991.

Myers JR, Kisner SM, Fosbroke DE, Lifetime Risk of Fatal Occupational Injuries within Industries, by Occupation, Gender, and Race. Human and Ecological Risk Assessment.(4), 6:1291-1307. 1998.

NIOSH, National Occupational Research Agenda (NORA). Center for Disease Control and Prevention (CDC) NIOSH. DHHS (NIOSH). Publication No. 98-141. Cincinnati. 1998.

Punnett L, Robins JM, Wegman DH, Keyserling WM. Soft tissue disorders in the upper limbs of female garment workers. Scand J Work Env Hlth, 11:417, 1985.

Punnett L. The costs of work-related musculoskeletal disorders in automotive manufacturing. New Solutions: A Journal of Environmental and Occupational Health Policy 2000, 9 (4): 403-426, 1999.

Quinn MM, Kriebel D, Geiser K, Moure-Eraso R. Sustainable Production: a proposed strategy for the work environment. Am J Ind Med. 34: 297-304. 1998.

Rom, WN. Environmental & Occupational Medicine, 3nd edition. Little, Brown & Co. 1998.

Rosenstock L, Cullen M. Clinical Occupational Medicine. WB Saunders Blue Book; 1996.

Rutstein DD, Mullan RJ, Frazier TM, Halperin WE, Melius JM, and Sestito JP. Sentinel health events (occupational): a basis for physician recognition and public health surveillance. Am J Public Health; 9:198-221;1983.

Silverstein BA, Stetson DS, Keyserling WM, Fine LJ. Work-related musculoskeletal disorders: Comparison of data sources for surveillance. Amer J Industr Med 31: 600-608, 1997.

Sokas RK, Spiegelman D, Wegman DH. Self-reported musculoskeletal complaints among garment workers. Am J Indust Med, 15:197-206, 1989.

U.S. Department of Labor, Bureau of Labor Statistics, Table 11. Employed Persons by Detailed Occupation, Sex, Race and Hispanic Origin. ftp://ftp.bls.gov/Pub/special.request/If/aat11.txt

Wallerstein NB, Weinger M, eds. Special Issue on Empowerment approaches to worker health and safety education. Am J Indust Med. 22(5), 619-783, 1992.

Wallerstein N, Rubinstein HL. Teaching about job hazards: A guide for workers and their health providers. American Public Health Association (APHA), 1993.

Wagner G. and Wegman DH. Comments on: Risk and Incidence of asthma attributable to occupational exposure among HMO members. Am J Ind Med. 35. 2: 206. 1999.